Internet of Things –
From Research and Innovation to
Market Deployment

RIVER PUBLISHERS SERIES IN COMMUNICATIONS

Consulting Series Editors

MARINA RUGGIERI
University of Roma "Tor Vergata"
Italy

HOMAYOUN NIKOOKAR
Delft University of Technology
The Netherlands

This series focuses on communications science and technology. This includes the theory and use of systems involving all terminals, computers, and information processors; wired and wireless networks; and network layouts, procontentsols, architectures, and implementations.

Furthermore, developments toward newmarket demands in systems, products, and technologies such as personal communications services, multimedia systems, enterprise networks, and optical communications systems.

- Wireless Communications
- Networks
- Security
- Antennas & Propagation
- Microwaves
- Software Defined Radio

For a list of other books in this series, visit www.riverpublishers.com
http://riverpublishers.com/river publisher/series.php?msg=Communications

Internet of Things –
From Research and Innovation to
Market Deployment

Editors

Dr. Ovidiu Vermesan

SINTEF, Norway

Dr. Peter Friess

EU, Belgium

LONDON AND NEW YORK

Published 2014 by River Publishers
River Publishers
Alsbjergvej 10, 9260 Gistrup, Denmark
www.riverpublishers.com

Distributed exclusively by Routledge
4 Park Square, Milton Park, Abingdon, Oxon OX14 4RN
605 Third Avenue, New York, NY 10158

First published in paperback 2024

Internet of Things – From Research and Innovation to Market Deployment / by Ovidiu Vermesan, Peter Friess.

Routledge is an imprint of the Taylor & Francis Group, an informa business

Publisher's Note
The publisher has gone to great lengths to ensure the quality of this reprint but points out that some imperfections in the original copies may be apparent.

While every effort is made to provide dependable information, the publisher, authors, and editors cannot be held responsible for any errors or omissions.

ISBN: 978-87-93102-94-1 (hbk)
ISBN: 978-87-7004-495-0 (pbk)
ISBN: 978-1-003-33862-8 (ebk)

DOI: 10.1201/9781003338628

Dedication

"Creativity is inventing, experimenting, growing, taking risks, breaking rules, making mistakes, and having fun."

— Mary Lou Cook

"Around here, however, we don't look backwards for very long. We keep moving forward, opening up new doors and doing new things, because we're curious... and curiosity keeps leading us down new paths."

— Walt Disney

Acknowledgement

The editors would like to thank the European Commission for their support in the planning and preparation of this book. The recommendations and opinions expressed in the book are those of the editors and contributors, and do not necessarily represent those of the European Commission.

Ovidiu Vermesan
Peter Friess

Contents

Preface

Shaping the Future of Internet of Things Applications

The potential benefits of Internet of Things (IoT) are almost limitless and IoT applications are changing the way we work and live by saving time and resources and opening new opportunities for growth, innovation and knowledge creation. The Internet of Things allows private and public-sector organizations to manage assets, optimize performance, and develop new business models. As a vital instrument to interconnect devices and to act as generic enabler of the hyper-connected society, the Internet of Things has great potential to support an ageing society, to improve the energy efficiency and to optimise all kinds of mobility and transport. The complementarity with approaches like cyber-physical systems, cloud technologies, big data and future networks like 5G is highly evident. The success of the Internet of Things will depend on the ecosystem development, supported by an appropriate regulatory environment and a climate of trust, where issues like identification, trust, privacy, security, and semantic interoperability are pivotal.

The following chapters will provide insights on the state-of-the-art of research and innovation in IoT and will expose you to the progress towards the deployment of Internet of Things applications.

Editors Biography

Dr. Ovidiu Vermesan holds a Ph.D. degree in microelectronics and a Master of International Business (MIB) degree. He is Chief Scientist at SINTEF Information and Communication Technology, Oslo, Norway. His research interests are in the area of microelectronics/nanoelectronics, analog and mixed-signal ASIC Design (CMOS/BiCMOS/SOI) with applications in measurement, instrumentation, high-temperature applications, medical electronics and integrated sensors; low power/low voltage ASIC design; and computer-based electronic analysis and simulation. Dr. Vermesan received SINTEFs 2003 award for research excellence for his work on the implementation of a biometric sensor system. He is currently working with projects addressing nanoelectronics integrated systems, communication and embed- ded systems, integrated sensors, wireless identifiable systems and RFID for future Internet of Things architectures with applications in green automotive, internet of energy, healthcare, oil and gas and energy efficiency in buildings. He has authored or co-authored over 75 technical articles and conference papers. He is actively involved in the activities of the new Electronic Components and Systems for European Leadership (ECSEL) Joint Technology Initiative (JTI). He coordinated and managed various national and international/EU projects related to integrated electronics. Dr. Vermesan is the coordinator of the IoT European Research Cluster (IERC) of the European Commission, actively participated in projects related to Internet of Things.

Dr. Peter Friess is a senior official of DG CONNECT of the European Commission, taking care for more than six years of the research and innovation policy for the Internet of Things. In his function he has shaped the on-going European research and innovation program on the Internet of Things and accompanied the European Commission's direct investment of 70 Mill. Euro in this field. He also oversees the international cooperation on the Internet of Things, in particular with Asian countries. In previous engagements he was working as senior consultant for IBM, dealing with major automotive and utility companies in Germany and Europe. Prior to this engagement he worked as

IT manager at Philips Semiconductors on with business process optimisation in complex manufacturing. Before this period he was active as researcher in European and national research projects on advanced telecommunications and business process reorganisation. He is a graduated engineer in Aeronautics and Space technology from the University of Munich and holds a Ph.D. in Systems Engineering including self-organising systems from the University of Bremen. He also published a number of articles and co-edits a yearly book of the European Internet of Things Research Cluster.

1

Introduction

Thibaut Kleiner

DG Connect, European Commission

Eighteen months ago, the emergence of the Internet of Things (IoT) was still considered with a certain degree of scepticism. These days are gone. A series of announcements, from the acquisition of Nest Labs by Google for $3.2 billion to Samsung Gear and health-related wearables to the development of Smart Home features into Apple's iOS, have made IoT an increasingly tangible business opportunity. Predictions have been consistently on the high side in terms of potential. For instance, Cisco estimates that the Internet of Things has a potential value of $14 trillion. Looking at the buzz in the US as well as in Asia, one may wonder whether it means that Europe has once more missed the technology train and that IoT will be developed by the likes of Apple, Google and Samsung. Or whether public research is still relevant given the fast moving market developments.

From the European Commission's point of view, it would be a serious mistake to believe that it is *game over* for IoT. In fact, the hope has been building for some years and we are only at the very beginning. The EU has already for some time invested in supporting Research and Innovation in the field of IoT, notably in the areas of embedded systems and cyber-physical systems, network technologies, semantic inter-operability, operating platforms and security, and generic enablers. Just like RFID did not quite manage to become pervasive yet, there are still a number of challenges before the IoT can expand and reach maturity. Research results are now feeding into innovation, and a series of components are now available, which could usefully be exploited and enhanced by the market. But there are still a number of issues as regards how Internet of Things applications will develop and be deployed on the back of Research and Innovation.

These issues may be of a technical nature, not least in terms of security, reliability, complex integration, discoverability and interoperability. Standardisation will certainly play a role there. Other issues may be related to the acceptability of IoT applications by users and by citizens. Others may relate to business models and generally to market partitioning and coordination problems, which could seriously hamper the deployment of IoT applications.

In that context, the Commission is considering how to best support IoT Research and Innovation further. One opportunity could be around pilot projects testing the deployment of large amounts of sensors in relation with Big Data applications. Another could be to launch large scale pilots to test in real life the possibility for integrated IoT solutions to be delivered. End-to-end security is another clear challenge that will need to be addressed to convince users to adopt the IoT.

Despite the hype around American and Asian mobile device manufacturers, IoT's research and technology is still very strong in Europe, and there are many examples of successful European companies. Europe has potentially a full eco-system with market leaders on smart sensors (Bosch, STMicroelectronics), embedded systems (ARM, Infineon), software (Atos, SAP), network vendors (Ericsson), telecoms (Orange) and application integrators (Siemens, Philips) or dynamic SMEs with huge growing potential (Zigpos, Libelium, Enevo) and industrial early adopters like BMW or Airbus. There is still hope that European players will emerge as the winners of the forthcoming IoT revolution. The EC will do its utmost to support that process. This book is a very useful contribution in that context and it shows that the Internet of Things European Research Cluster has been a driven force for the deployment of IoT not only in Europe, but globally.

2

Putting the Internet of Things Forward to the Next Nevel

Peter Friess and Francisco Ibanez

DG Connect, European Commission

2.1 The Internet of Things Today

The Internet of Things (IoT) is defined by ITU and IERC as a dynamic global network infrastructure with self-configuring capabilities based on standard and interoperable communication protocols where physical and virtual "things" have identities, physical attributes and virtual personalities, use intelligent interfaces and are seamlessly integrated into the information network. Over the last year, IoT has moved from being a futuristic vision - with sometimes a certain degree of hype - to an increasing market reality.

Significant business decisions have been taken by major ICT players like Google, Apple and Cisco to position themselves in the IoT landscape. Telecom operators consider that Machine-to-Machine (M2M) and the Internet of Things are becoming a core business focus, reporting significant growth in the number of connected objects in their networks. Device manufactures e.g. concerning wearable devices anticipate a full new business segment towards a wider adoption of the IoT.

The EU has already for some time invested in supporting Research and Innovation in the field of IoT, notably in the areas of embedded systems and cyber-physical systems, network technologies, semantic interoperability, operating platforms and security, and generic enablers. These research results are now feeding into innovation, and a series of components are available, which could usefully be exploited and enhanced by the market.

In line with this development, the majority of the governments in Europe, in Asia, and in the Americas consider the Internet of Things as an area of innovation and growth. Although larger players in some application areas still do not recognise the potential, many of them pay high attention or even accelerate the pace by coining new terms for the IoT and adding additional components to it. In addition end-users in the private and business domain have nowadays acquired a significant competence in dealing with smart devices and networked applications.

As the Internet of Things continues to develop, further potential is estimated by a combination with related technology approaches and concepts such as Cloud computing, Future Internet, Big Data, Robotics and Semantic technologies. The idea is of course not new as such but, as these concepts overlap in some parts (technical and service architectures, virtualisation, interoperability, automation), genuine innovators see more the aspect of complementarity rather than defending individual domains.

2.2 The Internet of Things Tomorrow

Not only the assimilation of ICT concepts and their constituencies are pivotal but also integrating them in smart environments and ecosystems across specific application domains. The overall challenge is to extend the current Internet of Things into a dynamically configured web of platforms for connected devices, objects, smart environments, services and persons.

Numerous industrial analyses (Acatech, Cisco, Ericsson, IDC, Forbes) have identified the evolution of the Internet of Things embedded in Smart Environments and Smart Platforms forming a smart web of everything as one of the next big concepts to support societal changes and economic growth, which will support the citizen in their professional and domestic/public life. By the end of the decade, dozens of connected devices per human being on the planet are conservatively anticipated, relating to a business whose yearly growth is estimated at 20%. In this context Europe needs to maintain its position through leadership in smart and embedded systems technologies with a strong potential in the evolving market of cyber-physical systems.

On the way towards "Platforms for Connected Smart Objects" the biggest challenge will be to overcome the fragmentation of vertically-oriented closed systems and architectures and application areas towards open systems and integrated environments and platforms, which support multiple applications of social value by bringing contextual knowledge of the surrounding world and events into complex business/social processes. The task is to create and master innovative ecosystems beyond smart phones and device markets. Play from multiple application sectors including potential new players, which do not exist today exist are called upon to play a role in such an endeavour.

In order to specify challenges for IoT relating to deployment, technological and business model validation and acceptability large-scale pilots could play an important role, addressing security and trust issues in an integrated manner, and contributing to certification and validation ecosystems in the IoT arena. These pilots would appropriately fit with the objectives called for in the European Innovation Partnership for Smart Cities, eHealth and in the Electronics Leaders Group. An additional opportunity has been identified in sharing IoT large-scale pilots' approaches and results with China, Japan, Korea and the US.

A non-exhaustive list of objectives for IoT large-scale pilots would address the following topics:

- **Solving remaining technological barriers**, with a strong focus on security. From an industrial perspective, European technology providers could be leading such pilots. In addition, remaining engineering issues need to be solved, speeding up the engineering process for conceiving, designing, testing and validating IoT based systems. Relating to software aspects, it is important to manage a very high number of IoT devices that cannot be controlled individually but need be run automatically.
- **Exploring the integration potential** of IoT architectures and components together with Cloud solutions and Big Data approaches, as this conceptual novel approach needs to be substantiated in depth. Moreover, the actors in the fields are still continuing to develop and exploit their own domains, be it IoT, Cloud or Big Data.
- **Validating user acceptability**, focusing on applications, which are not operational today, and still do require some research. One such example could be car-to-car communication or enhanced assisted living for the purpose of relaying safety critical information. Those kinds of applications also come with regulatory issues, e.g. in terms of liability.
- **Promoting innovation on sensor/object platforms.** The Future Internet pilot activities have fostered this type of pilots by giving the power to a set of users in order to develop innovative applications out of data that are collected from the sensors. More innovation is certainly also needed in the way non-experienced users could communicate with smart objects.
- **Demonstrating cross use cases issues**, to validate the concepts of generic technologies that can serve a multiplicity of environments

and imply the cooperation of incumbents, like e.g. for Smart Homes, Smart manufacturing, dedicated Smart City areas, Smart Food Value Chain or Digital social communities, creative industries, city and regional development. In addition it is essential to run pilots deploying agent-driven applications and to test system of systems in physical spaces in relation to the human scale.

2.3 Potential Success Factors

The Internet of Things Technologies will foster European core industrial activities such as industry automation, generation and distribution of renewable energies (Smart Grid), as well as the development and production of enhanced environmental technologies, cars, airplanes, etc. The future IoT will be a cornerstone for the development of smart and sustainable cities and smart and sustainable infrastructures in general.

Key success factors for promising differentiation of the European IoT Technology players can be formulated as follows for technological, user concerned, business and societal aspects:

- Mitigation of architecture/system divergences through a common architecture framework for connected system qualities and interoperability
- Development of IoT technologies that support the shift from data collection to knowledge creation
- Focus on IoT Value Chain development and adequate analysis from the start of product development towards user acceptance
- Development of a legal framework to ensure adequate consideration of trust and ethical issues

This article expresses the personal view of the authors and in no way constitutes a formal or official position of the European Commission.

3

Internet of Things Strategic Research and Innovation Agenda

Ovidiu Vermesan[1], Peter Friess[2], Patrick Guillemin[3], Harald Sundmaeker[4], Markus Eisenhauer[5], Klaus Moessner[6], Marilyn Arndt[7], Maurizio Spirito[8], Paolo Medagliani[9], Raffaele Giaffreda[10], Sergio Gusmeroli[11], Latif Ladid[12], Martin Serrano[13], Manfred Hauswirth[13], Gianmarco Baldini[14]

[1] *SINTEF, Norway*
[2] *European Commission, Belgium*
[3] *ETSI, France*
[4] *ATB GmbH, Germany*
[5] *Fraunofer FIT, Germany*
[6] *University of Surrey, UK*
[7] *Orange, France*
[8] *ISMB, Italy*
[9] *Thales Communications & Security, France*
[10] *CREATE-NET, Italy*
[11] *TXT e-solutions, Italy*
[12] *University of Luxembourg, Luxembourg*
[13] *Digital Enterprise Research Institute, Galway, Ireland*
[14] *Joint Research Centre, European Commission, Italy*

"Whatever you can do, or dream you can, begin it. Boldness has genius, power and magic in it."

Johann Wolfgang von Goethe

"If you want something new, you have to stop doing something old."

Peter F. Drucker

"Vision is the art of seeing things invisible."

Jonathan Swift

7

3.1 Internet of Things Vision

Internet of Things (IoT) is a concept and a paradigm that considers pervasive presence in the environment of a variety of things/objects that through wireless and wired connections and unique addressing schemes are able to interact with each other and cooperate with other things/objects to create new applications/services and reach common goals. In this context the research and development challenges to create a smart world are enormous. A world where the real, digital and the virtual are converging to create smart environments that make energy, transport, cities and many other areas more intelligent. The goal of the Internet of Things is to enable things to be connected anytime, anyplace, with anything and anyone ideally using any path/network and any service. Internet of Things is a new revolution of the Internet. Objects make themselves recognizable and they obtain intelligence by making or enabling context related decisions thanks to the fact that they can communicate information about themselves and they can access information that has been aggregated by other things, or they can be components of complex services [69].

The Internet of Things is the network of physical objects that contain embedded technology to communicate and sense or interact with their internal states or the external environment and the confluence of efficient wireless protocols, improved sensors, cheaper processors, and a bevy of start-ups and established companies developing the necessary management and application software has finally made the concept of the Internet of Things mainstream. The number of Internet-connected devices surpassed the number of human beings on the planet in 2011, and by 2020, Internet-connected devices are expected to number between 26 billion and 50 billion. For every Internet-connected PC or handset there will be 5–10 other types of devices sold with native Internet connectivity [43].

According to industry analyst firm IDC, the installed base for the Internet of Things will grow to approximately 212 billion devices by 2020, a number that includes 30 billion connected devices. IDC sees this growth driven largely by intelligent systems that will be installed and collecting data - across both consumer and enterprise applications [44].

These types of applications can involve the electric vehicle and the smart house, in which appliances and services that provide notifications, security, energy-saving, automation, telecommunication, computers and entertainment will be integrated into a single ecosystem with a shared user interface. IoT is providing access to information, media and services, through wired and

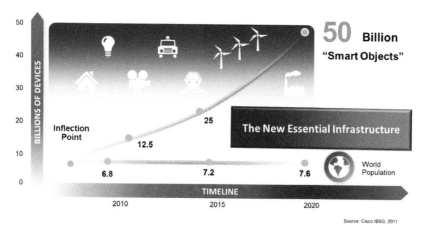

Figure 3.1 Internet-connected devices and the future evolution (Source: Cisco, 2011)

wireless broadband connections. The Internet of Things makes use of synergies that are generated by the convergence of Consumer, Business and Industrial Internet Consumer, Business and Industrial Internet. The convergence creates the open, global network connecting people, data, and things. This convergence leverages the cloud to connect intelligent things that sense and transmit a broad array of data, helping creating services that would not be obvious without this level of connectivity and analytical intelligence. The use of platforms is being driven by transformative technologies such as cloud, things, and mobile. The Internet of Things and Services makes it possible to create networks incorporating the entire manufacturing process that convert factories into a smart environment. The cloud enables a global infrastructure to generate new services, allowing anyone to create content and applications for global users. Networks of things connect things globally and maintain their identity online. Mobile allows connection to this global infrastructure anytime, anywhere. The result is a globally accessible network of things, users, and consumers, who are available to create businesses, contribute content, generate and purchase new services.

Platforms also rely on the power of network effects, as they allow more things, they become more valuable to the other things and to users that make use of the services generated. The success of a platform strategy for IoT can be determined by connection, attractiveness and knowledge/information/ data flow.

The European Commission while recognizing the potential of Converging Sciences and Technologies Converging Sciences and Technologies to advance

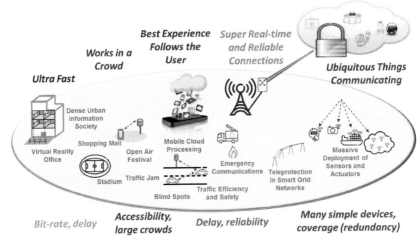

Figure 3.2   Future Communication Challenges – 5G scenarios [2]

the Lisbon Agenda, proposes a bottom-up approach to prioritize the setting of a particular goal for convergence of science and technology research; meet challenges and opportunities for research and governance and allow for integration of technological potential as well as recognition of limits, European needs, economic opportunities, and scientific interests.

Enabling technologies for the Internet of Things considered in [36] can be grouped into three categories: *i*) technologies that enable "things" to acquire contextual information, *ii*) technologies that enable "things" to process contextual information, and *iii*) technologies to improve security and privacy. The first two categories can be jointly understood as functional building blocks required building "intelligence" into "things", which are indeed the features that differentiate the IoT from the usual Internet. The third category is not a *functional* but rather a *de facto* requirement, without which the penetration of the IoT would be severely reduced. Internet of Things developments implies that the environments, cities, buildings, vehicles, clothing, portable devices and other objects have more and more information associated with them and/or the ability to sense, communicate, network and produce new information. In addition the network technologies have to cope with the new challenges such as very high data rates, dense crowds of users, low latency, low energy, low cost and a massive number of devices, The 5G scenarios that reflect the future challenges and will serve as guidance for further work are outlined by the EC funded METIS project [2].

As the Internet of Things becomes established in smart factories, both the volume and the level of detail of the corporate data generated will increase. Moreover, business models will no longer involve just one company, but will instead comprise highly dynamic networks of companies and completely new value chains. Data will be generated and transmitted autonomously by smart machines and these data will inevitably cross company boundaries. A number of specific dangers are associated with this new context – for example, data that were initially generated and exchanged in order to coordinate manufacturing and logistics activities between different companies could, if read in conjunction with other data, suddenly provide third parties with highly sensitive information about one of the partner companies that might, for example, give them an insight into its business strategies. New instruments will be required if companies wish to pursue the conventional strategy of keeping such knowledge secret in order to protect their competitive advantage. New, regulated business models will also be necessary – the raw data that are generated may contain information that is valuable to third parties and companies may therefore wish to make a charge for sharing them. Innovative business models like this will also require legal safeguards (predominantly in the shape of contracts) in order to ensure that the value added created is shared out fairly, e.g. through the use of dynamic pricing models [55].

3.1.1 Internet of Things Common Definition

Ten "critical" trends and technologies impacting IT for the next five years were laid out by Gartner and among them the Internet of Things. All of these things have an IP address and can be tracked. The Internet is expanding into enterprise assets and consumer items such as cars and televisions. The problem is that most enterprises and technology vendors have yet to explore the possibilities of an expanded Internet and are not operationally or organizationally ready. Gartner [54] identifies four basic usage models that are emerging:

- Manage
- Monetize
- Operate
- Extend.

These can be applied to people, things, information, and places, and therefore the so called "Internet of Things" will be succeeded by the "Internet of Everything."

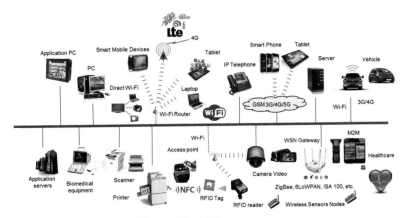

Figure 3.3   IP Convergence

In this context the notion of network convergence using IP is fundamental and relies on the use of a common multi-service IP network supporting a wide range of applications and services.

The use of IP to communicate with and control small devices and sensors opens the way for the convergence of large, IT-oriented networks with real time and specialized networked applications.

The fundamental characteristics of the IoT are as follows [65]:

- Interconnectivity: With regard to the IoT, anything can be interconnected with the global information and communication infrastructure.
- Things-related services: The IoT is capable of providing thing-related services within the constraints of things, such as privacy protection and semantic consistency between physical things and their associated virtual things. In order to provide thing-related services within the constraints of things, both the technologies in physical world and information world will change.
- Heterogeneity: The devices in the IoT are heterogeneous as based on different hardware platforms and networks. They can interact with other devices or service platforms through different networks.
- Dynamic changes: The state of devices change dynamically, e.g., sleeping and waking up, connected and/or disconnected as well as the context of devices including location and speed. Moreover, the number of devices can change dynamically.
- Enormous scale: The number of devices that need to be managed and that communicate with each other will be at least an order of magnitude

larger than the devices connected to the current Internet. The ratio of communication triggered by devices as compared to communication triggered by humans will noticeably shift towards device-triggered communication. Even more critical will be the management of the data generated and their interpretation for application purposes. This relates to semantics of data, as well as efficient data handling.

The Internet of Things is not a single technology, it's a concept in which most new things are connected and enabled such as street lights being networked and things like embedded sensors, image recognition functionality, augmented reality, near field communication are integrated into situational decision support, asset management and new services. These bring many business opportunities and add to the complexity of IT [52].

To accommodate the diversity of the IoT, there is a heterogeneous mix of communication technologies, which need to be adapted in order to address the needs of IoT applications such as energy efficiency, security, and reliability. In this context, it is possible that the level of diversity will be scaled to a number a manageable connectivity technologies that address the needs of the IoT applications, are adopted by the market, they have already proved to be serviceable, supported by a strong technology alliance. Examples of standards in these categories include wired and wireless technologies like Ethernet, Wi-Fi, Bluetooth, ZigBee, and Z-Wave.

Distribution, transportation, logistics, reverse logistics, field service, etc. are areas where the coupling of information and "things" may create new business processes or may make the existing ones highly efficient and more profitable.

The Internet of Things provides solutions based on the integration of information technology, which refers to hardware and software used to store, retrieve, and process data and communications technology which includes electronic systems used for communication between individuals or groups. The rapid convergence of information and communications technology is taking place at three layers of technology innovation: the cloud, data and communication pipes/networks and device [46].

The synergy of the access and potential data exchange opens huge new possibilities for IoT applications. Already over 50% of Internet connections are between or with things. In 2011 there were over 15 billion things on the Web, with 50 billion+ intermittent connections.

By 2020, over 30 billion connected things, with over 200 billion with intermittent connections are forecast. Key technologies here include

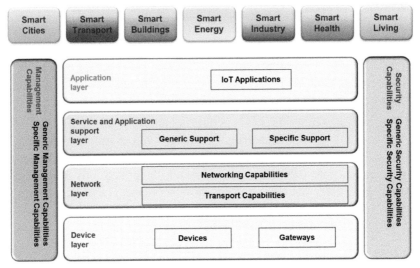

Figure 3.4   IoT Layered Architecture (Source: ITU-T)

embedded sensors, image recognition and NFC. By 2015, in more than 70% of enterprises, a single executable will oversee all Internet connected things. This becomes the Internet of Everything [53].

As a result of this convergence, the IoT applications require that classical industries are adapting and the technology will create opportunities for new industries to emerge and to deliver enriched and new user experiences and services.

In addition, to be able to handle the sheer number of things and objects that will be connected in the IoT, cognitive technologies and contextual intelligence are crucial. This also applies for the development of context aware applications that need to be reaching to the edges of the network through smart devices that are incorporated into our everyday life.

The Internet is not only a network of computers, but it has evolved into a network of devices of all types and sizes, vehicles, smartphones, home appliances, toys, cameras, medical instruments and industrial systems, all connected, all communicating and sharing information all the time.

The Internet of Things had until recently different means at different levels of abstractions through the value chain, from lower level semiconductor through the service providers.

The Internet of Things is a "global concept" and requires a common definition. Considering the wide background and required technologies,

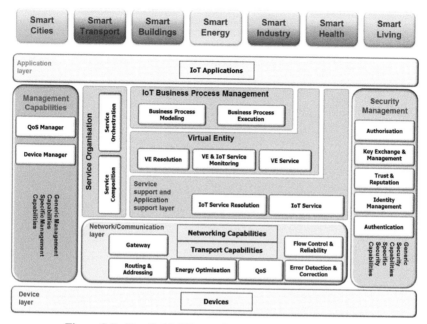

Figure 3.5 Detailed IoT Layered Architecture (Source: IERC)

from sensing device, communication subsystem, data aggregation and pre-processing to the object instantiation and finally service provision, generating an unambiguous definition of the "Internet of Things" is non-trivial.

The IERC is actively involved in ITU-T Study Group 13, which leads the work of the International Telecommunications Union (ITU) on standards for next generation networks (NGN) and future networks and has been part of the team which has formulated the following definition [65]: *"**Internet of things (IoT)**: A global infrastructure for the information society, enabling advanced services by interconnecting (physical and virtual) things based on existing and evolving interoperable information and communication technologies. NOTE 1 – Through the exploitation of identification, data capture, processing and communication capabilities, the IoT makes full use of things to offer services to all kinds of applications, whilst ensuring that security and privacy requirements are fulfilled. NOTE 2 – From a broader perspective, the IoT can be perceived as a vision with technological and societal implications."*

The IERC definition [67] states that IoT is *"A dynamic global network infrastructure with self-configuring capabilities based on standard and inter-operable communication protocols where physical and virtual "things" have*

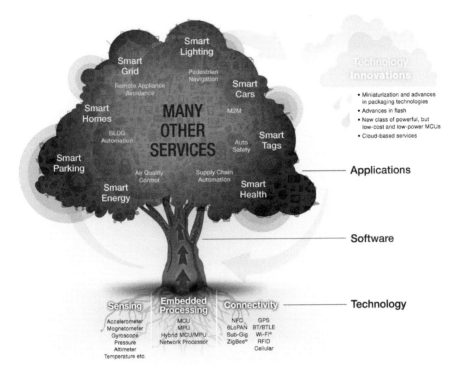

Figure 3.6   The IoT: Different Services, Technologies, Meanings for Everyone [77]

identities, physical attributes, and virtual personalities and use intelligent interfaces, and are seamlessly integrated into the information network.".

3.2 IoT Strategic Research and Innovation Directions

The development of enabling technologies such as nanoelectronics, communications, sensors, smart phones, embedded systems, cloud networking, network virtualization and software will be essential to provide to things the capability to be connected all the time everywhere. This will also support important future IoT product innovations affecting many different industrial sectors. Some of these technologies such as embedded or cyber-physical systems form the edges of the Internet of Things bridging the gap between cyber space and the physical world of real things, and are crucial in enabling the Internet of Things to deliver on its vision and become part of bigger systems in a world of "systems of systems".

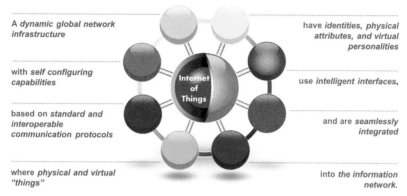

A dynamic global network infrastructure

with *self configuring* capabilities

based on *standard and interoperable* communication protocols

where *physical and virtual* "things"

have *identities, physical attributes, and virtual personalities*

use *intelligent interfaces,*

and are *seamlessly integrated*

into *the information network.*

Figure 3.7　IoT Definition [68]

The final report of the Key Enabling Technologies (KET), of the High-Level Expert Group [47] identified the enabling technologies, crucial to many of the existing and future value chains of the European economy:

- Nanotechnologies.
- Micro and Nano electronics
- Photonics
- Biotechnology
- Advanced Materials
- Advanced Manufacturing Systems.

As such, IoT creates intelligent applications that are based on the supporting KETs identified, as IoT applications address smart environments either physical or at cyber-space level, and in real time.

To this list of key enablers, we can add the global deployment of IPv6 across the World enabling a global and ubiquitous addressing of any communicating smart thing.

From a technology perspective, the continuous increase in the integration density proposed by Moore's Law was made possible by a dimensional scaling: in reducing the critical dimensions while keeping the electrical field constant, one obtained at the same time a higher speed and a reduced power consumption of a digital MOS circuit: these two parameters became driving forces of the microelectronics industry along with the integration density.

The International Technology Roadmap for Semiconductors has emphasized in its early editions the "miniaturization" and its associated benefits in terms of performances, the traditional parameters in Moore's Law. This trend for increased performances will continue, while performance can always be traded against power depending on the individual application, sustained by the incorporation into devices of new materials, and the application of new transistor concepts. This direction for further progress is labelled "More Moore".

The second trend is characterized by functional diversification of semiconductor-based devices. These non-digital functionalities do contribute to the miniaturization of electronic systems, although they do not necessarily scale at the same rate as the one that describes the development of digital functionality. Consequently, in view of added functionality, this trend may be designated "More-than-Moore" [50].

Mobile data traffic is projected to double each year between now and 2015 and mobile operators will find it increasingly difficult to provide the bandwidth requested by customers. In many countries there is no additional spectrum that can be assigned and the spectral efficiency of mobile networks is reaching its physical limits. Proposed solutions are the seamless integration of existing Wi-Fi networks into the mobile ecosystem. This will have a direct impact on Internet of Things ecosystems.

The chips designed to accomplish this integration are known as "multi-com" chips. Wi-Fi and baseband communications are expected to converge and the architecture of mobile devices is likely to change and the baseband chip is expected to take control of the routing so the connectivity components are connected to the baseband or integrated in a single silicon package. As a result of this architecture change, an increasing share of the integration work is likely done by baseband manufacturers (ultra -low power solutions) rather than by handset producers.

The market for wireless communications is one of the fastest-growing segments in the integrated circuit industry. Breath takingly fast innovation, rapid changes in communications standards, the entry of new players, and the evolution of new market sub segments will lead to disruptions across the industry. LTE and multicom solutions increase the pressure for industry consolidation, while the choice between the ARM and x86 architectures forces players to make big bets that may or may not pay off [63].

Integrated networking, information processing, sensing and actuation capabilities allow physical devices to operate in changing environments. Tightly coupled cyber and physical systems that exhibit high level of integrated intelligence are referred to as cyber-physical systems. These systems are part of the enabling technologies for Internet of Things applications where computational and physical processes of such systems are tightly interconnected and coordinated to work together effectively, with or without the humans in the

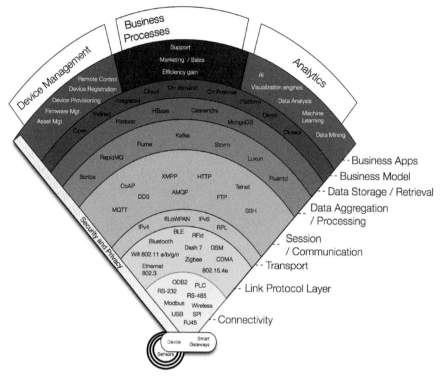

Figure 3.8 IoT landscape [21]

loop. Robots, intelligent buildings, implantable medical devices, vehicles that drive themselves or planes that automatically fly in a controlled airspace, are examples of cyber-physical systems that could be part of Internet of Things ecosystems.

Today many European projects and initiatives address Internet of Things technologies and knowledge. Given the fact that these topics can be highly diverse and specialized, there is a strong need for integration of the individual results. Knowledge integration, in this context is conceptualized as the process through which disparate, specialized knowledge located in multiple projects across Europe is combined, applied and assimilated.

The Strategic Research and Innovation Agenda (SRIA) is the result of a discussion involving the projects and stakeholders involved in the IERC activities, which gather the major players of the European ICT landscape addressing IoT technology priorities that are crucial for the competitiveness of European industry:

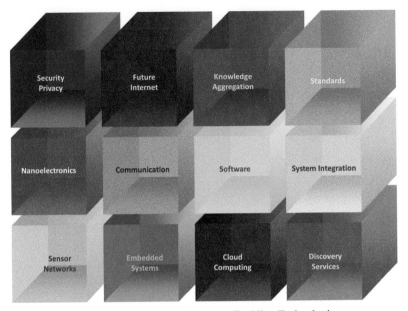

Figure 3.9   Internet of Things — Enabling Technologies

IERC Strategic Research and Innovation Agenda covers the important issues and challenges for the Internet of Things technology. It provides the vision and the roadmap for coordinating and rationalizing current and future research and development efforts in this field, by addressing the different enabling technologies covered by the Internet of Things concept and paradigm.

Many other technologies are converging to support and enable IoT applications. These technologies are summarised as:

- IoT architecture
- Identification
- Communication
- Networks technology
- Network discovery
- Software and algorithms
- Hardware technology
- Data and signal processing
- Discovery and search engine
- Network management
- Power and energy storage
- Security, trust, dependability and privacy

- Interoperability
- Standardization

The Strategic Research and Innovation Agenda is developed with the support of a European-led community of interrelated projects and their stakeholders, dedicated to the innovation, creation, development and use of the Internet of Things technology.

Since the release of the first version of the Strategic Research and Innovation Agenda, we have witnessed active research on several IoT topics. On the one hand this research filled several of the gaps originally identified in the Strategic Research and Innovation Agenda, whilst on the other it created new challenges and research questions. Recent advances in areas such as cloud computing, cyber-physical systems, autonomic computing, and social networks have changed the scope of the Internet of Thing's convergence even more so. The Cluster has a goal to provide an updated document each year that records the relevant changes and illustrates emerging challenges. The updated release of this Strategic Research and Innovation Agenda builds incrementally on previous versions [68], [69], [84], [85], [85] and highlights the main research topics that are associated with the development of IoT enabling technologies, infrastructures and applications with an outlook towards 2020 [73].

The research items introduced will pave the way for innovative applications and services that address the major economic and societal challenges underlined in the EU 2020 Digital Agenda [74].

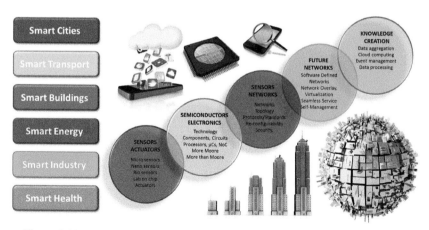

Figure 3.10 Internet of Things - Smart Environments and Smart Spaces Creation

The IERC Strategic Research and Innovation Agenda is developed incrementally based on its previous versions and focus on the new challenges being identified in the last period.

The timeline of the Internet of Things Strategic Research and Innovation Agenda covers the current decade with respect to research and the following years with respect to implementation of the research results. Of course, as the Internet and its current key applications show, we anticipate unexpected trends will emerge leading to unforeseen and unexpected development paths.

The Cluster has involved experts working in industry, research and academia to provide their vision on IoT research challenges, enabling technologies and the key applications, which are expected to arise from the current vision of the Internet of Things.

The IoT Strategic Research and Innovation Agenda covers in a logical manner the vision, the technological trends, the applications, the technology enablers, the research agenda, timelines, priorities, and finally summarises in two tables the future technological developments and research needs.

Advances in embedded sensors, processing and wireless connectivity are bringing the power of the digital world to objects and places in the physical world. IoT Strategic Research and Innovation Agenda is aligned with the findings of the 2011 Hype Cycle developed by Gartner [76], which includes the broad trend of the Internet of Things, called the "real-world Web" in earlier Gartner research.

The field of the Internet of Things is based on the paradigm of supporting the IP protocol to all edges of the Internet and on the fact that at the edge of the network many (very) small devices are still unable to support IP protocol stacks. This means that solutions centred on minimum Internet of Things devices are considered as an additional Internet of Things paradigm *without IP to all access edges*, due to their importance for the development of the field.

3.2.1 IoT Applications and Use Case Scenarios

The IERC vision is that "the major objectives for IoT are the creation of smart environments/spaces and self-aware things (for example: smart transport, products, cities, buildings, rural areas, energy, health, living, etc.) for climate, food, energy, mobility, digital society and health applications"[68].

The outlook for the future is the emerging of a network of interconnected uniquely identifiable objects and their virtual representations in an Internet alike structure that is positioned over a network of interconnected computers allowing for the creation of a new platform for economic growth.

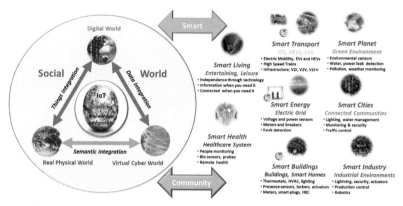

Figure 3.11 Internet of Things in the context of Smart Environments and Applications [84]

Smart is the new green as defined by Frost & Sullivan [51] and the green products and services will be replaced by smart products and services. Smart products have a real business case, can typically provide energy and efficiency savings of up to 30 per cent, and generally deliver a two- to three-year return on investment. This trend will help the deployment of Internet of Things applications and the creation of smart environments and spaces.

At the city level, the integration of technology and quicker data analysis will lead to a more coordinated and effective civil response to security and safety (law enforcement and blue light services); higher demand for outsourcing security capabilities.

At the building level, security technology will be integrated into systems and deliver a return on investment to the end-user through leveraging the technology in multiple applications (HR and time and attendance, customer behaviour in retail applications etc.).

There will be an increase in the development of "Smart" vehicles which have low (and possibly zero) emissions. They will also be connected to infrastructure. Additionally, auto manufacturers will adopt more use of "Smart" materials.

The key focus will be to make the city smarter by optimizing resources, feeding its inhabitants by urban farming, reducing traffic congestion, providing more services to allow for faster travel between home and various destinations, and increasing accessibility for essential services. It will become essential to have intelligent security systems to be implemented at key junctions in the city. Various types of sensors will have to be used to make this a reality. Sensors are moving from "smart" to "intelligent". Biometrics is already integrated in the smart mobile phones and is expected to be used together with CCTV at highly sensitive locations around the city. National identification cards will also become an essential tool for the identification of an individual. In addition, smart cities in 2020 will require real time auto identification security systems.

The IoT brings about a paradigm were everything is connected and will redefine the way humans and machines interface and the way they interact with the world around them.

Fleet Management is used to track vehicle location, hard stops, rapid acceleration, and sudden turns using sophisticated analysis of the data in order to implement new policies (e.g., no right/left turns) that result in cost savings for the business.

Today there are billions of connected sensors already deployed with smart phones and many other sensors are connected to these smart mobile network using different communication protocols.

The challenges is in getting the data from them in an interoperable format and in creating systems that break vertical silos and harvest the data across domains, thus unleashing truly useful IoT applications that are user centred, context aware and create new services by communication across the verticals.

Wastewater treatment plants will evolve into bio-refineries. New, innovative wastewater treatment processes will enable water recovery to help close the growing gap between water supply and demand.

Self-sensing controls and devices will mark new innovations in the Building Technologies space. Customers will demand more automated, self-controlled solutions with built in fault detection and diagnostic capabilities.

Development of smart implantable chips that can monitor and report individual health status periodically will see rapid growth.

Smart pumps and smart appliances/devices are expected to be significant contributors towards efficiency improvement. Process equipment with in built "smartness" to self-assess and generate reports on their performance, enabling efficient asset management, will be adopted.

Test and measurement equipment is expected to become smarter in the future in response to the demand for modular instruments having lower power consumption. Furthermore, electronics manufacturing factories will become more sustainable with renewable energy and sell unused energy back to the grid, improved water conservation with rain harvesting and implement other smart building technologies, thus making their sites "Intelligent Manufacturing Facilities".

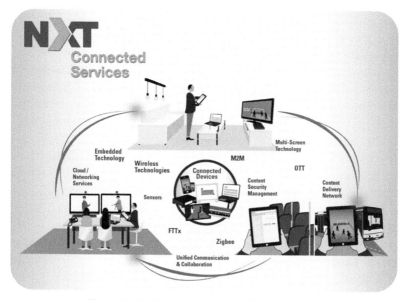

Figure 3.12 Connected Devices Illustration [62]

General Electric Co. considers that this is taking place through the convergence of the global industrial system with the power of advanced computing, analytics, low-cost sensing and new levels of connectivity permitted by the Internet. The deeper meshing of the digital world with the world of machines holds the potential to bring about profound transformation to global industry, and in turn to many aspects of daily life [58].

The Industrial Internet starts with embedding sensors and other advanced instrumentation in an array of machines from the simple to the highly complex. This allows the collection and analysis of an enormous amount of data, which can be used to improve machine performance, and inevitably the efficiency of the systems and networks that link them. Even the data itself can become "intelligent," instantly knowing which users it needs to reach.

Consumer IoT is essentially wireless, while the industrial IoT has to deal with an installed base of millions of devices that could potentially become part of this network (many legacy systems installed before IP deployment). These industrial objects are linked by wires that provides the reliable communications needed. The industrial IoT has to consider the legacy using specialised protocols, including Lonworks, DeviceNet, Profibus and CAN and they will be connected into this new netwoek of networks through gateways.

The automation and management of asset-intensive enterprises will be transformed by the rise of the IoT, Industry 4.0, or simply Industrial Internet. Compared with the Internet revolution, many product and asset management solutions have labored under high costs and poor connectivity and performance. This is now changing. New high-performance systems that can support both Internet and Cloud connectivity as well as predictive asset management are reaching the market. New cloud computing models, analytics, and aggregation technologies enable broader and low cost application of analytics across these much more transparent assets. These developments have the potential to radically transform products, channels, and company business models. This will create disruptions in the business and opportunities for all types of organizations - OEMs, technology suppliers, system integrators, and global consultancies. There may be the opportunity to overturn established business models, with a view toward answering customer pain points and also growing the market in segments that cannot be served economically with today's offerings. Mobility, local diagnostics, and remote asset monitoring are important components of these new solutions, as all market participants need ubiquitous access to their assets, applications, and customers. Real-time mobile applications support EAM, MRO, inventory management, inspections, workforce management, shop floor interactions, facilities management, field service automation, fleet management, sales and marketing, machine-to-machine (M2M), and many others [56]

In this context the new concept of Internet of Energy requires web based architectures to readily guarantee information delivery on demand and to change the traditional power system into a networked Smart Grid that is largely automated, by applying greater intelligence to operate, enforce policies, monitor and self-heal when necessary. This requires the integration and interfacing of the power grid to the network of data represented by the Internet, embracing energy generation, transmission, delivery, substations, distribution control, metering and billing, diagnostics, and information systems to work seamlessly and consistently.

This concept would enable the ability to produce, store and efficiently use energy, while balancing the supply/demand by using a cognitive Internet of Energy that harmonizes the energy grid by processing the data, information and knowledge via the Internet. The Internet of Energy concept as presented in Figure 3.14 [35] will leverage on the information highway provided by the Internet to link devices and services with the distributed smart energy grid that is the highway for renewable energy resources allowing stakeholders to

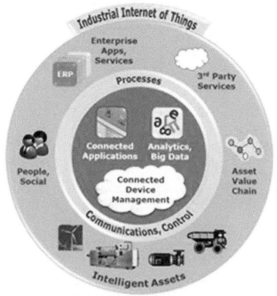

Industrial Internet of Things (IoT)
Enables New Business Models

Figure 3.13 Industrial Internet of Things [56]

use green technologies and sell excess energy back to the utility. The concept has the energy management element in the centre of the communication and exchange of data and energy.

The Internet of Energy applications are connected through the Future Internet and "Internet of Things" enabling seamless and secure interactions and cooperation of intelligent embedded systems over heterogeneous communication infrastructures.

It is expected that this "development of smart entities will encourage development of the novel technologies needed to address the emerging challenges of public health, aging population, environmental protection and climate change, conservation of energy and scarce materials, enhancements to safety and security and the continuation and growth of economic prosperity." The IoT applications are further linked with Green ICT, as the IoT will drive energy-efficient applications such as smart grid, connected electric cars, energy-efficient buildings, thus eventually helping in building green intelligent cities.

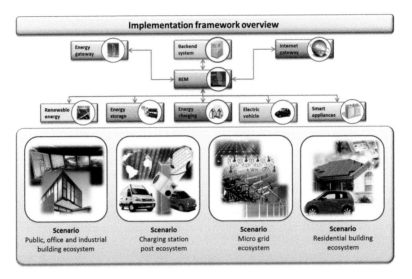

Figure 3.14   Internet of Energy Implementation Framework (Source:[35])

3.2.2 IoT Functional View

The Internet of Things concept refers to uniquely identifiable things with their virtual representations in an Internet-like structure and IoT solutions comprising a number of components such as:

- Module for interaction with local IoT devices (for example embedded in a mobile phone or located in the immediate vicinity of the user and thus contactable via a short range wireless interface). This module is responsible for acquisition of observations and their forwarding to remote servers for analysis and permanent storage.
- Module for local analysis and processing of observations acquired by IoT devices.
- Module for interaction with remote IoT devices, directly over the Internet or more likely via a proxy. This module is responsible for acquisition of observations and their forwarding to remote servers for analysis and permanent storage.
- Module for application specific data analysis and processing. This module is running on an application server serving all clients. It is taking requests from mobile and web clients and relevant IoT observations as input, executes appropriate data processing algorithms and generates output in terms of knowledge that is later presented to users.

- Module for integration of IoT-generated information into the business processes of an enterprise. This module will be gaining importance with the increased use of IoT data by enterprises as one of the important factors in day-to-day business or business strategy definition.
- User interface (web or mobile): visual representation of measurements in a given context (for example on a map) and interaction with the user, i.e. definition of user queries.

It is important to highlight that one of the crucial factors for the success of IoT is stepping away from vertically-oriented, closed systems towards open systems, based on open APIs and standardized protocols at various system levels.

In this context innovative architecture and platforms are needed to support highly complex and inter-connected IoT applications. A key consideration is how to enable development and application of comprehensive architectural frameworks that include both the physical and cyber elements based on enabling technologies. In addition considering the technology convergence trend new platforms will be needed for communication and to effectively extract actionable information from vast amounts of raw data, while providing a robust timing and systems framework to support the real-time control and synchronization requirements of complex, networked, engineered physical/cyber/virtual systems.

A large number of applications made available through application markets have significantly helped the success of the smart phone industry. The development of such a huge number of smart phone applications is primarily due to involvement of the developers' community at large. Developers leveraged smart phone open platforms and the corresponding development tools, to create a variety of applications and to easily offer them to a growing number of users through the application markets.

Similarly, an IoT ecosystem has to be established, defining open APIs for developers and offering appropriate channels for delivery of new applications. Such open APIs are of particular importance on the level of the module for application specific data analysis and processing, thus allowing application developers to leverage the underlying communication infrastructure and use and combine information generated by various IoT devices to produce new, added value.

Although this might be the most obvious level at which it is important to have open APIs, it is equally important to aim towards having such APIs defined on all levels in the system. At the same time one should have in mind the heterogeneity and diversity of the IoT application space. This will truly support the development of an IoT ecosystem that encourages development of new applications and new business models.

The complete system will have to include supporting tools providing security and business mechanisms to enable interaction between a numbers of different business entities that might exist [86].

Research challenges:

- Design of open APIs on all levels of the IoT ecosystem
- Design of standardized formats for description of data generated by IoT devices to allow mashups of data coming from different domains and/or providers.

3.2.3 Application Areas

In the last few years the evolution of markets and applications, and therefore their economic potential and their impact in addressing societal trends and challenges for the next decades has changed dramatically. Societal trends are grouped as: health and wellness, transport and mobility, security and safety, energy and environment, communication and e-society. These trends create significant opportunities in the markets of consumer electronics, automotive electronics, medical applications, communication, etc. The applications in in these areas benefit directly by the More-Moore and More-than-Moore semiconductor technologies, communications, networks and software developments.

Potential applications of the IoT are numerous and diverse, permeating into practically all areas of every-day life of individuals, enterprises, and society as a whole. The IERC [68–69], [84–85] has identified and described the main Internet of Things applications, which span numerous applications domains: smart energy, smart health, smart buildings, smart transport, smart industry and smart city. The vision of a pervasive IoT requires the integration of the various domains into a single, unified, domain and addresses the enabling technologies needed for these domains while taking into account the elements that form the third dimension like security, privacy, trust, safety.

The IoT application domains identified by IERC [68], [85] are based on inputs from experts, surveys [86] and reports [87]. The IoT application covers "smart" environments/spaces in domains such as: Transportation, Building, City, Lifestyle, Retail, Agriculture, Factory, Supply chain, Emergency, Health care, User interaction, Culture and tourism, Environment and Energy.

The applications areas include as well the domain of Industrial Internet [58] where intelligent devices, intelligent systems, and intelligent decision-making

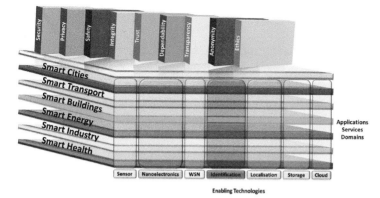

Figure 3.15 IoT 3D Matrix

represent the primary ways in which the physical world of machines, facilities, fleets and networks can more deeply merge with the connectivity, big data and analytics of the digital world. Manufacturing and industrial automation are under pressure from shortened product life-cycles and the demand for a shorter time to market in many areas. The next generation of manufacturing systems will therefore be built with flexibility and reconfiguration as a fundamental objective.

This change is eminent in the transition from traditional, centralized control applications to an interconnected, cooperative "Internet of Things" model. Strong hierarchies are broken in favour of meshed, networks and formerly passive devices are replaced with "smart objects" that are network enabled and can perform compute operations. The software side has to match and leverage the changes in the hardware. Service Oriented Architectures (SOAs) are a well-known concept from business computing to deal with flexibility and reconfiguration requirements in a loosely coupled manner. However, the common concepts of SOAs cannot be directly mapped to embedded networks and industrial control applications, because of the hard boundary conditions, such as limited resources and real-time requirements [57].

The updated list of IoT applications presented below, includes examples of IoT applications in different domains, which is showing why the Internet of Things is one of the strategic technology trends for the next 5 years.

Smart Food/Water Monitoring
Water Quality: Study of water suitability in rivers and the sea for fauna and eligibility for drinkable use.

Water Leakages: Detection of liquid presence outside tanks and pressure variations along pipes.

River Floods: Monitoring of water level variations in rivers, dams and reservoirs.

Water Management: Real-time information about water usage and the status of waterlines could be collected by connecting residential water meters to an Internet protocol (IP) network. As a consequence could be reductions in labour and maintenance costs, improved accuracy and lower costs in meter readings, and possibly water consumption reductions.

Supply Chain Control: Monitoring of storage conditions along the supply chain and product tracking for traceability purposes.

Wine Quality Enhancing: Monitoring soil moisture and trunk diameter in vineyards to control the amount of sugar in grapes and grapevine health.

Green Houses: Control micro-climate conditions to maximize the production of fruits and vegetables and its quality.

Golf Courses: Selective irrigation in dry zones to reduce the water resources required in the green.

In-field Monitoring: Reducing spoilage and food waste with better monitoring, statistic handling, accurate ongoing data obtaining, and management of the agriculture fields, including better control of fertilizing, electricity and watering.

Smart Health

Fall Detection: Assistance for elderly or disabled people living independent.

Physical Activity Monitoring for Aging People: Body sensors network measures motion, vital signs, unobtrusiveness and a mobile unit collects, visualizes and records activity data.

Medical Fridges: Control of conditions inside freezers storing vaccines, medicines and organic elements.

Sportsmen Care: Vital signs monitoring in high performance centres and fields. Health and fitness products for these purposes exist, that measure exercise, steps, sleep, weight, blood pressure, and other statistics.

Patients Surveillance: Monitoring of conditions of patients inside hospitals and in old people's home.

Chronic Disease Management: Patient-monitoring systems with comprehensive patient statistics could be available for remote residential monitoring of patients with chronic diseases such as pulmonary and heart diseases and diabetes. The reduced medical center admissions, lower costs, and shorter hospital stays would be some of the benefits.

Ultraviolet Radiation: Measurement of UV sun rays to warn people not to be exposed in certain hours.

Hygienic hand control: RFID-based monitoring system of wrist bands in combination of Bluetooth LE tags on a patient's doorway controlling hand hygiene in hospitals, where vibration notifications is sent out to inform about time for hand wash; and all the data collected produce analytics which can be used to potentially trace patient infections to particular healthcare workers.

Sleep control: Wireless sensors placed across the mattress sensing small motions, like breathing and heart rate and large motions caused by tossing and turning during sleep, providing data available through an app on the smartphone.

Dental Health: Bluetooth connected toothbrush with smartphone app analyzes the brushing uses and gives information on the brushing habits on the smartphone for private information or for showing statistics to the dentist.

Smart Living

Intelligent Shopping Applications: Getting advice at the point of sale according to customer habits, preferences, presence of allergic components for them, or expiring dates.

Energy and Water Use: Energy and water supply consumption monitoring to obtain advice on how to save cost and resources. Maximizing energy efficiency by introducing lighting and heating products, such as bulbs, thermostats and air conditioners.

Remote Control Appliances: Switching on and off remotely appliances to avoid accidents and save energy.

Weather Station: Displays outdoor weather conditions such as humidity, temperature, barometric pressure, wind speed and rain levels using meters with ability to transmit data over long distances.

Smart Home Appliances: Refrigerators with LCD screen telling what's inside, food that's about to expire, ingredients you need to buy and with all the information available on a smartphone app. Washing machines allowing you to monitor the laundry remotely, and run automatically when electricity rates are lowest. Kitchen ranges with interface to a smartphone app allowing remotely adjustable temperature control and monitoring the oven's self-cleaning feature.

Gas Monitoring: Real-information about gas usage and the status of gas lines could be provided by connecting residential gas meters to an Internet protocol (IP) network. As for the water monitoring, the possible outcome could be reductions in labor and maintenance costs, improved accuracy and lower costs in meter readings, and possibly gas consumption reductions.

Safety Monitoring: Baby monitoring, cameras, and home alarm systems making people feel safe in their daily life at home.

Smart Jewelry: Increased personal safety by wearing a piece of jewelry inserted with Bluetooth enabled technology used in a way that a simple push establishes contact with your smartphone, which through an app will send alarms to selected people in your social circle with information that you need help and your location.

Smart Environment Monitoring

Forest Fire Detection: Monitoring of combustion gases and preemptive fire conditions to define alert zones.

Air Pollution: Control of CO_2 emissions of factories, pollution emitted by cars and toxic gases generated in farms.

Landslide and Avalanche Prevention: Monitoring of soil moisture, vibrations and earth density to detect dangerous patterns in land conditions.

Earthquake Early Detection: Distributed control in specific places of tremors.

Protecting wildlife: Tracking collars utilizing GPS/GSM modules to locate and track wild animals and communicate their coordinates via SMS.

Meteorological Station Network: Study of weather conditions in fields to forecast ice formation, rain, drought, snow or wind changes.

Marine and Coastal Surveillance: Using different kinds of sensors integrated in planes, unmanned aerial vehicles, satellites, ship etc. to control the maritime activities and traffic in important areas, keep track of fishing boats, supervise environmental conditions and dangerous oil cargo etc.

Smart Manufacturing

Smart Product Management: Control of rotation of products in shelves and warehouses to automate restocking processes.

Compost: Control of humidity and temperature levels in alfalfa, hay, straw, etc. to prevent fungus and other microbial contaminants.

Offspring Care: Control of growing conditions of the offspring in animal farms to ensure its survival and health.

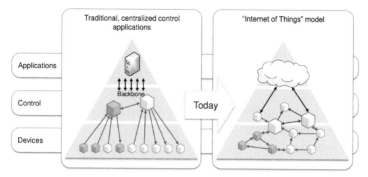

Figure 3.16 Interconnected, Cooperative "Internet of Things" Model for Manufacturing and Industrial Automation [57]

Animal Tracking: Location and identification of animals grazing in open pastures or location in big stables.

Toxic Gas Levels: Study of ventilation and air quality in farms and detection of harmful gases from excrements.

Production Line: Monitoring and management of the production line using RFID, sensors, video monitoring, remote information distribution and cloud solutions enabling the production line data to be transferred to the enterprise-based systems. This may result in more quickly improvement of the entire product quality assurance process by decision makers, updated workflow charts, and inspection procedures delivered to the proper worker groups via digital displays in real time.

Telework: Offering the employees technologies that enable home offices would reduce costs, improve productivity, and add employment opportunities at the same time as reducing real estate for employees, lower office maintenance and cleanings, and eliminating daily office commute.

Smart Energy

Smart Grid: Energy consumption monitoring and management.

Photovoltaic Installations: Monitoring and optimization of performance in solar energy plants.

Wind Turbines: Monitoring and analyzing the flow of energy from wind turbines, and two-way communication with consumers' smart meters to analyze consumption patterns.

Water Flow: Measurement of water pressure in water transportation systems.

Radiation Levels: Distributed measurement of radiation levels in nuclear power stations surroundings to generate leakage alerts.

Power Supply Controllers: Controller for AC-DC power supplies that determines required energy, and improve energy efficiency with less energy waste for power supplies related to computers, telecommunications, and consumer electronics applications.

Smart Buildings

Perimeter Access Control: Access control to restricted areas and detection of people in non-authorized areas.

Liquid Presence: Liquid detection in data centres, warehouses and sensitive building grounds to prevent break downs and corrosion.

Indoor Climate Control: Measurement and control of temperature, lighting, CO_2 fresh air in ppm etc.

Intelligent Thermostat: Thermostat that learns the users programming schedule after a few days, and from that programs itself. Can be used with an app to connect to the thermostat from a smart telephone, where control, watching the energy history, how much energy is saved and why can be displayed.

Intelligent Fire Alarm: System with sensors measuring smoke and carbon monoxide, giving both early warnings, howling alarms and speaks with a human voice telling where the smoke is or when carbon monoxide levels are rising, in addition to giving a message on the smartphone or tablet if the smoke or CO alarm goes off.

Intrusion Detection Systems: Detection of window and door openings and violations to prevent intruders.

Motion Detection: Infrared motion sensors which reliably sends alerts to alarm panel (or dialer) and with a system implementing reduced false alarms algorithms and adaption to environmental disturbances.

Art and Goods Preservation: Monitoring of conditions inside museums and art warehouses.

Residential Irrigation: Monitoring and smart watering system.

Smart Transport and Mobility

NFC Payment: Payment processing based in location or activity duration for public transport, gyms, theme parks, etc.

Quality of Shipment Conditions: Monitoring of vibrations, strokes, container openings or cold chain maintenance for insurance purposes.

Item Location: Searching of individual items in big surfaces like warehouses or harbours.

Storage Incompatibility Detection: Warning emission on containers storing inflammable goods closed to others containing explosive material.

Fleet Tracking: Control of routes followed for delicate goods like medical drugs, jewels or dangerous merchandises.

Electric Vehicle Charging Stations Reservation: Locates the nearest charging station and tell the user whether its in use. Drivers can ease their range anxiety by reserving charging stations ahead of time. Help the planning of extended EV road trips, so the EV drivers make the most of potential charging windows

Vehicle Auto-diagnosis: Information collection from CAN Bus to send real time alarms to emergencies or provide advice to drivers.

Management of cars: Car sharing companies manages the use of vehicles using the Internet and mobile phones through connections installed in each car.

Road Pricing: Automatic vehicle payment systems would improve traffic conditions and generate steady revenues if such payments are introduced in busy traffic zones. Reductions in traffic congestions and reduced CO_2 emissions would be some of the benefits.

Connected Militarized Defence: By connecting command-centre facilities, vehicles, tents, and Special Forces real-time situational awareness for combat personnel in war areas and visualization of the location of allied/enemy personnel and material would be provided.

Smart Industry

Tank level: Monitoring of water, oil and gas levels in storage tanks and cisterns.

Silos Stock Calculation: Measurement of emptiness level and weight of the goods.

Explosive and Hazardous Gases: Detection of gas levels and leakages in industrial environments, surroundings of chemical factories and inside mines. Meters can transmit data that will be reliably read over long distances.

M2M Applications: Machine auto-diagnosis and assets control.

Maintenance and repair: Early predictions on equipment malfunctions and service maintenance can be automatically scheduled ahead of an actual part failure by installing sensors inside equipment to monitor and send reports.

Indoor Air Quality: Monitoring of toxic gas and oxygen levels inside chemical plants to ensure workers and goods safety.

Temperature Monitoring: Control of temperature inside industrial and medical fridges with sensitive merchandise.

Ozone Presence: Monitoring of ozone levels during the drying meat process in food factories.

Indoor Location: Asset indoor location by using active (ZigBee, UWB) and passive tags (RFID/NFC).

Aquaculture industry monitoring: Remotely operating and monitoring operational routines on the aquaculture site, using sensors, cameras, wireless communication infrastructure between sites and land base, winch systems etc. to perform site and environment surveillance, feeding and system operations.

Smart City

Smart Parking: Real-time monitoring of parking spaces availability in the city making residents able to identify and reserve the closest available spaces. Reduction in traffic congestions and increased revenue from dynamic pricing could be some of the benefits as well as simpler responsibility for traffic wardens recognizing non-compliant usage.

Structural Health: Monitoring of vibrations and material conditions in buildings, bridges and historical monuments.

Noise Urban Maps: Sound monitoring in bar areas and centric zones in real time.

Traffic Congestion: Monitoring of vehicles and pedestrian levels to optimize driving and walking routes.

Smart Lightning: Intelligent and weather adaptive lighting in street lights.

Waste Management: Detection of rubbish levels in containers to optimize the trash collection routes. Garbage cans and recycle bins with RFID tags allow the sanitation staff to see when garbage has been put out. Maybe "Pay as you throw"-programs would help to decrease garbage waste and increase recycling efforts.

Intelligent Transportation Systems: Smart Roads and Intelligent Highways with warning messages and diversions according to climate conditions and unexpected events like accidents or traffic jams.

Safe City: Digital video monitoring, fire control management, public announcement systems

Connected Learning: Improvements in teacher utilization, reduction in instructional supplies, productivity improvement, and lower costs are examples of benefits that may be gained from letting electronic resources deliver data-driven, authentic and collaborative learning experience to larger groups.

Smart irrigation of public spaces: Maintenance of parks and lawns by burying park irrigation monitoring sensors in the ground wirelessly connected to repeaters and with a wireless gateway connection to Internet.

Smart Tourism: Smartphone Apps supported by QR codes and NFC tags providing interesting and useful tourist information throughout the city. The information could include museums, art galleries, libraries, touristic attractions, tourism offices, monuments, shops, buses, taxis, gardens, etc.

The IoT application space is very diverse and IoT applications serve different users. Different user categories have different driving needs. From the IoT perspective there are three important user categories:

- The individual citizens
- Community of citizens (citizens of a city, a region, country or society as a whole)
- The enterprises.

Examples of the individual citizens/human users' needs for the IoT applications are as follows:

- To increase their safety or the safety of their family members - for example remotely controlled alarm systems, or activity detection for elderly people;
- To make it possible to execute certain activities in a more convenient manner - for example: a personal inventory reminder;

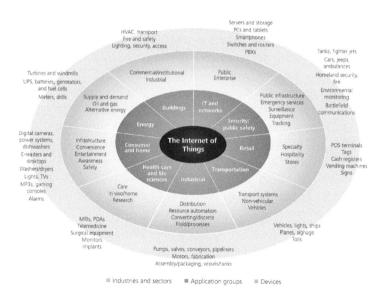

Figure 3.17 Internet of Things- proliferation of connected devices across industries (Source: Beecham Research, [75])

- To generally improve life-style - for example monitoring health parameters during a workout and obtaining expert's advice based on the findings, or getting support during shopping;
- To decrease the cost of living - for example building automation that will reduce energy consumption and thus the overall cost.

The society as a user has different drivers. It is concerned with issues of importance for the whole community, often related to medium to longer term challenges.

Some of the needs driving the society as a potential user of IoT are the following:

- To ensure public safety - in the light of various recent disasters such as the nuclear catastrophe in Japan, the tsunami in the Indian Ocean, earthquakes, terrorist attacks, etc. One of the crucial concerns of the society is to be able to predict such events as far ahead as possible and to make rescue missions and recovery as efficient as possible. One good example of an application of IoT technology was during the Japan nuclear catastrophe, when numerous Geiger counters owned by individuals were connected to the Internet to provide a detailed view of radiation levels across Japan.
- To protect the environment

    o Requirements for reduction of carbon emissions have been included in various legislations and agreements aimed at reducing the impact on the planet and making sustainable development possible.
    o Monitoring of various pollutants in the environment, in particular in the air and in the water.
    o Waste management, not just general waste, but also electrical devices and various dangerous goods are important and challenging topics in every society.
    o Efficient utilization of various energy and natural resources are important for the development of a country and the protection of its resources.

- To create new jobs and ensure existing ones are sustainable - these are important issues required to maintain a high level quality of living.

Enterprises, as the third category of IoT users have different needs and different drivers that can potentially push the introduction of IoT-based solutions.

Examples of the needs are as follows:

- Increased productivity - this is at the core of most enterprises and affects the success and profitability of the enterprise;
- Market differentiation - in a market saturated with similar products and solutions, it is important to differentiate, and IoT is one of the possible differentiators;
- Cost efficiency - reducing the cost of running a business is a "mantra" for most of the CEOs. Better utilization of resources, better information used in the decision process or reduced downtime are some of the possible ways to achieve this.

The explanations of the needs of each of these three categories are given from a European perspective. To gain full understanding of these issues, it is important to capture and analyse how these needs are changing across the world. With such a complete picture, we will be able to drive IoT developments in the right direction.

Another important topic which needs to be understood is the business rationale behind each application. In other words, understanding the value an application creates.

Important research questions are: who takes the cost of creating that value; what are the revenue models and incentives for participating, using or contributing to an application? Again due to the diversity of the IoT application domain and different driving forces behind different applications, it will not be possible to define a universal business model. For example, in the case of applications used by individuals, it can be as straightforward as charging a fee for a service, which will improve their quality of life. On the other hand, community services are more difficult as they are fulfilling needs of a larger community. While it is possible that the community as a whole will be willing to pay (through municipal budgets), we have to recognise the limitations in public budgets, and other possible ways of deploying and running such services have to be investigated.

3.3 IoT Smart-X Applications

It is impossible to envisage all potential IoT applications having in mind the development of technology and the diverse needs of potential users. In the following sections, we present several applications, which are important. These applications are described, and the research challenges are identified. The IoT applications are addressing the societal needs and the advancements to enabling technologies such as nanoelectronics and cyber-physical systems continue to be challenged by a variety of technical (i.e., scientific and engineering), institutional, and economical issues.

The list is focusing to the applications chosen by the IERC as priorities for the next years and it provides the research challenges for these applications. While the applications themselves might be different, the research challenges are often the same or similar.

3.3.1 Smart Cities

By 2020 we will see the development of Mega city corridors and networked, integrated and branded cities. With more than 60 percent of the world population expected to live in urban cities by 2025, urbanization as a trend will have diverging impacts and influences on future personal lives and mobility. Rapid expansion of city borders, driven by increase in population and infrastructure development, would force city borders to expand outward and engulf the surrounding daughter cities to form mega cities, each with a population of more than 10 million. By 2023, there will be 30 mega cities globally, with 55 percent in developing economies of India, China, Russia and Latin America [51].

This will lead to the evolution of smart cities with eight smart features, including Smart Economy, Smart Buildings, Smart Mobility, Smart Energy, Smart Information Communication and Technology, Smart Planning, Smart Citizen and Smart Governance. There will be about 40 smart cities globally by 2025.

The role of the cities governments will be crucial for IoT deployment. Running of the day-to-day city operations and creation of city development strategies will drive the use of the IoT. Therefore, cities and their services represent an almost ideal platform for IoT research, taking into account city requirements and transferring them to solutions enabled by IoT technology.

In Europe, the largest smart city initiatives completely focused on IoT is undertaken by the FP7 SmartSantander project [69]. This project aims at deploying an IoT infrastructure comprising thousands of IoT devices spread across several cities (Santander, Guildford, Luebeck and Belgrade). This will enable simultaneous development and evaluation of services and execution of various research experiments, thus facilitating the creation of a smart city environment.

Similarly, the OUTSMART [88] project, one of the FI PPP projects, is focusing on utilities and environment in the cities and addressing the role of IoT in waste and water management, public lighting and transport systems as well as environment monitoring.

A vision of the smart city as "horizontal domain" is proposed by the BUTLER project [90], in which many vertical scenarios are integrated and concur to enable the concept of smart life.

A smart city is defined as a city that monitors and integrates conditions of all of its critical infrastructures, including roads, bridges, tunnels, rail/subways, airports, seaports, communications, water, power, even major buildings, can better optimize its resources, plan its preventive maintenance activities, and monitor security aspects while maximizing services to its citizens. Emergency response management to both natural as well as man-made challenges to the system can be focused. With advanced monitoring systems and built-in smart sensors, data can be collected and evaluated in real time, enhancing city management's decision-making. For example, resources can be committed prior to a water main break, salt spreading crews dispatched only when a specific bridge has icing conditions, and use of inspectors reduced by knowing condition of life of all structures. In the long term Smart Cities vision, systems and structures will monitor their own conditions and carry out self-repair, as needed. The physical environment, air, water, and surrounding green spaces will be monitored in non-obtrusive ways for optimal quality, thus creating an enhanced living and working environment that is clean, efficient, and secure and that offers these advantages within the framework of the most effective use of all resources [81].

An illustrative example is depicted in Figure 3.18 [96]. The deployment of ICT to create 'smart cities' is gaining momentum in Europe, according to a study by Frost & Sullivan, accentuated by the numerous pilot projects running at regional, country and EU levels. Initiatives revolve around energy and water efficiency, mobility, infrastructure and platforms for open cities, citizen involvement, and public administration services. They are co-funded by the European Union through its ICT Policy Support and 7th Framework programmes, but, the report says, there is no clear business model for the uptake of smart cities. Projects are carried out in the form of collaborative networks established between the research community, businesses, the public sector, citizens and the wider community, and they foster an open innovation approach. Technologies such as smart metering, wireless sensor networks, open platforms, high-speed broadband and cloud computing are key building blocks of the smart city infrastructure [96].

A smart city is a developed urban area that creates sustainable economic development and high quality of life by excelling in multiple key areas: economy, mobility, environment, people, living, and government [97].

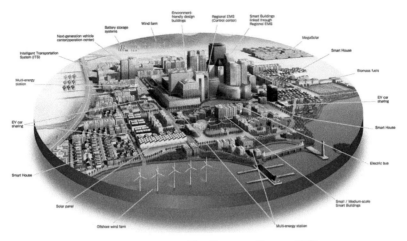

Figure 3.18   Smart City Concept. (Source: [95])

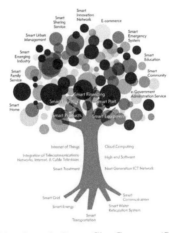

Figure 3.19   Organic Smart City Concept. (Source: [96])

Excelling in these key areas can be done so through strong human capital, social capital, and/or ICT infrastructure. With the introduction of IoT a city will act more like a living organism, a city that can respond to citizen's needs.

In this context there are numerous important research challenges for smart city IoT applications:

- Overcoming traditional silo based organization of the cities, with each utility responsible for their own closed world. Although not technological this is one of the main barriers

- Creating algorithms and schemes to describe information created by sensors in different applications to enable useful exchange of information between different city services
- Mechanisms for cost efficient deployment and even more important maintenance of such installations, including energy scavenging
- Ensuring reliable readings from a plethora of sensors and efficient calibration of a large number of sensors deployed everywhere from lampposts to waste bins
- Low energy protocols and algorithms
- Algorithms for analysis and processing of data acquired in the city and making "sense" out of it.
- IoT large scale deployment and integration

3.3.2 Smart Energy and the Smart Grid

There is increasing public awareness about the changing paradigm of our policy in energy supply, consumption and infrastructure. For several reasons our future energy supply should no longer be based on fossil resources. Neither is nuclear energy a future proof option. In consequence future energy supply needs to be based largely on various renewable resources. Increasingly focus must be directed to our energy consumption behaviour. Because of its volatile nature such supply demands an intelligent and flexible electrical grid which is able to react to power fluctuations by controlling electrical energy sources (generation, storage) and sinks (load, storage) and by suitable reconfiguration. Such functions will be based on networked intelligent devices (appliances, micro-generation equipment, infrastructure, consumer products) and grid infrastructure elements, largely based on IoT concepts. Although this ideally requires insight into the instantaneous energy consumption of individual loads (e.g. devices, appliances or industrial equipment) information about energy usage on a per-customer level is a suitable first approach.

Future energy grids are characterized by a high number of distributed small and medium sized energy sources and power plants which may be combined virtually ad hoc to virtual power plants; moreover in the case of energy outages or disasters certain areas may be isolated from the grid and supplied from within by internal energy sources such as photovoltaics on the roofs, block heat and power plants or energy storages of a residential area ("islanding").

A grand challenge for enabling technologies such as cyber-physical systems is the design and deployment of an energy system infrastructure that is able to provide blackout free electricity generation and distribution, is flexible enough to allow heterogeneous energy supply to or withdrawal from the grid,

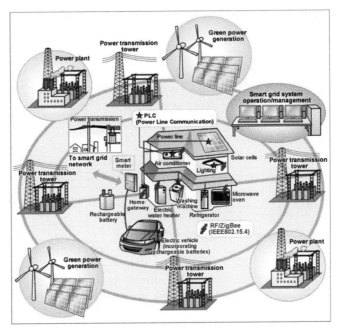

Figure 3.20   Smart Grid Representation

and is impervious to accidental or intentional manipulations. Integration of cyber-physical systems engineering and technology to the existing electric grid and other utility systems is a challenge. The increased system complexity poses technical challenges that must be considered as the system is operated in ways that were not intended when the infrastructure was originally built. As technologies and systems are incorporated, security remains a paramount concern to lower system vulnerability and protect stakeholder data [83]. These challenges will need to be address as well by the IoT applications that integrate heterogeneous cyber-physical systems.

The developing Smart Grid is expected to implement a new concept of transmission network which is able to efficiently route the energy which is produced from both concentrated and distributed plants to the final user with high security and quality of supply standards. Therefore the Smart Grid is expected to be the implementation of a kind of "Internet" in which the energy packet is managed similarly to the data packet - across routers and gateways which autonomously can decide the best pathway for the packet to reach its destination with the best integrity levels. In this respect the "Internet of Energy" concept is defined as a network infrastructure based on standard and

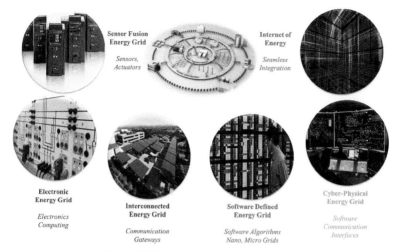

Figure 3.21 Internet of Energy Concept

interoperable communication transceivers, gateways and protocols that will allow a real time balance between the local and the global generation and storage capability with the energy demand. This will also allow a high level of consumer awareness and involvement.

The Internet of Energy (IoE) provides an innovative concept for power distribution, energy storage, grid monitoring and communication. It will allow units of energy to be transferred when and where it is needed. Power consumption monitoring will be performed on all levels, from local individual devices up to national and international level [102].

Saving energy based on an improved user awareness of momentary energy consumption is another pillar of future energy management concepts. Smart meters can give information about the instantaneous energy consumption to the user, thus allowing for identification and elimination of energy wasting devices and for providing hints for optimizing individual energy consumption. In a smart grid scenario energy consumption will be manipulated by a volatile energy price which again is based on the momentary demand (acquired by smart meters) and the available amount of energy and renewable energy production. In a virtual energy marketplace software agents may negotiate energy prices and place energy orders to energy companies. It is already recognised that these decisions need to consider environmental information such as weather forecasts, local and seasonal conditions. These must be to a much finer time scale and spatial resolution.

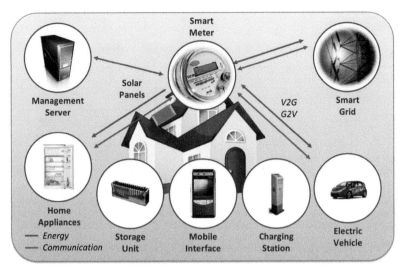

Figure 3.22   Internet of Energy: Residential Building Ecosystem [102]

In the long run electro mobility will become another important element of smart power grids. Electric vehicles (EVs) might act as a power load as well as moveable energy storage linked as IoT elements to the energy information grid (smart grid). IoT enabled smart grid control may need to consider energy demand and offerings in the residential areas and along the major roads based on traffic forecast. EVs will be able to act as sink or source of energy based on their charge status, usage schedule and energy price which again may depend on abundance of (renewable) energy in the grid. This is the touch point from where the following telematics IoT scenarios will merge with smart grid IoT.

This scenario is based on the existence of an IoT network of a vast multitude of intelligent sensors and actuators which are able to communicate safely and reliably. Latencies are critical when talking about electrical control loops. Even though not being a critical feature, low energy dissipation should be mandatory. In order to facilitate interaction between different vendors' products the technology should be based on a standardized communication protocol stack. When dealing with a critical part of the public infrastructure, data security is of the highest importance. In order to satisfy the extremely high requirements on reliability of energy grids, the components as well as their interaction must feature the highest reliability performance.

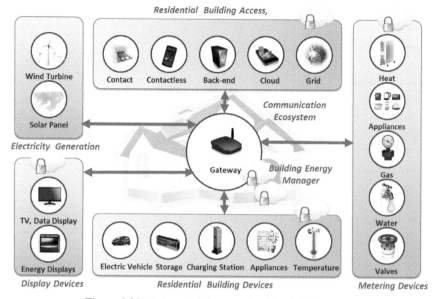

Figure 3.23 Internet of Energy – Residential Ecosystem

New organizational and learning strategies for sensor networks will be needed in order to cope with the shortcomings of classical hierarchical control concepts. The intelligence of smart systems does not necessarily need to be built into the devices at the systems' edges. Depending on connectivity, cloud-based IoT concepts might be advantageous when considering energy dissipation and hardware effort. Many IoT applications will go beyond one industrial sector. Energy, mobility and home/buildings sectors will share data through energy gateways that will control the transfer of energy and information.

Sophisticated and flexible data filtering, data mining and processing procedures and systems will become necessary in order to handle the high amount of raw data provided by billions of data sources. System and data models need to support the design of flexible systems which guarantee a reliable and secure real-time operation.

Some Research Challenges:

- Absolutely safe and secure communication with elements at the network edge
- Addressing scalability and standards interoperability
- Energy saving robust and reliable smart sensors/actuators

- Technologies for data anonymity addressing privacy concerns
- Dealing with critical latencies, e.g. in control loops
- System partitioning (local/cloud based intelligence)
- Mass data processing, filtering and mining; avoid flooding of communication network
- Real-time Models and design methods describing reliable interworking of heterogeneous systems (e.g. technical / economical / social / environmental systems). Identifying and monitoring critical system elements. Detecting critical overall system states in due time
- System concepts which support self-healing and containment of damage; strategies for failure contingency management
- Scalability of security functions
- Power grids have to be able to react correctly and quickly to fluctuations in the supply of electricity from renewable energy sources such as wind and solar facilities.

3.3.3 Smart Mobility and Transport

The connection of vehicles to the Internet gives rise to a wealth of new possibilities and applications which bring new functionalities to the individuals and/or the making of transport easier and safer. In this context the concept of Internet of Vehicles (IoV) [102] connected with the concept of Internet of Energy (IoE) represent future trends for smart transportation and mobility applications.

At the same time creating new mobile ecosystems based on trust, security and convenience to mobile/contactless services and transportation applications will ensure security, mobility and convenience to consumer-centric transactions and services.

Representing human behaviour in the design, development, and operation of cyber physical systems in autonomous vehicles is a challenge. Incorporating human-in-the-loop considerations is critical to safety, dependability, and predictability. There is currently limited understanding of how driver behaviour will be affected by adaptive traffic control cyber physical systems. In addition, it is difficult to account for the stochastic effects of the human driver in a mixed traffic environment (i.e., human and autonomous vehicle drivers) such as that found in traffic control cyber physical systems. Increasing integration calls for security measures that are not physical, but more logical while still ensuring there will be no security compromise. As cyber physical systems become more complex and interactions between components increases, safety and security

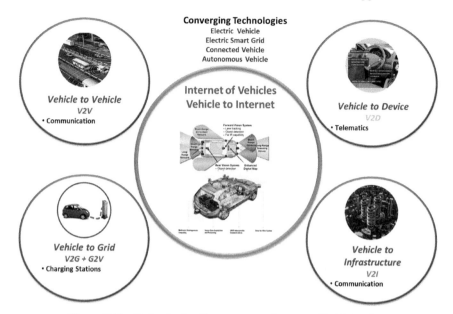

Figure 3.24 Technologies Convergence – Internet of Vehicles Case

will continue to be of paramount importance [83]. All these elements are of the paramount importance for the IoT ecosystems developed based on these enabling technologies.

When talking about IoT in the context of automotive and telematics, we may refer to the following application scenarios:

- Standards must be defined regarding the charging voltage of the power electronics, and a decision needs to be made as to whether the recharging processes should be controlled by a system within the vehicle or one installed at the charging station.
- Components for bidirectional operations and flexible billing for electricity need to be developed if electric vehicles are to be used as electricity storage media.
- **IoT as an inherent part of the vehicle control and management system**: Already today certain technical functions of the vehicles' on-board systems can be monitored on line by the service centre or garage to allow for preventative maintenance, remote diagnostics, instantaneous support and timely availability of spare parts. For this purpose data from on-board sensors are collected by a smart on-board unit and communicated via the Internet to the service centre.

- **IoT enabling traffic management and control**: Cars should be able to organise themselves in order to avoid traffic jams and to optimise drive energy usage. This may be done in coordination and cooperation with the infrastructure of a smart city's traffic control and management system. Additionally dynamic road pricing and parking tax can be important elements of such a system. Further mutual communications between the vehicles and with the infrastructure enable new methods for considerably increasing traffic safety, thus contributing to the reduction in the number of traffic accidents.

- **IoT enabling new transport scenarios (multi-modal transport)**: In such scenarios, e.g. automotive OEMs see themselves as mobility providers rather than manufacturers of vehicles. The user will be offered an optimal solution for transportation from A to B, based on all available and suitable transport means. Thus, based on the momentary traffic situation an ideal solution may be a mix of individual vehicles, vehicle sharing, railway, and commuter systems. In order to allow for seamless usage and on-time availability of these elements (including parking space), availability needs to be verified and guaranteed by online reservation and online booking, ideally in interplay with the above mentioned smart city traffic management systems.

- **Autonomous driving and interfacing with the infrastructure (V2V, V2I)**: The challenges address the interaction between the vehicle and the environment (sensors, actuators, communication, processing, information exchange, etc.) by considering road navigation systems that combines road localization and road shape estimation to drive on roads where a priori road geometry both is and is not available. Address a mixed-mode planning system that is able to both efficiently navigate on roads and safely manoeuvre through open areas and parking lots and develop a behavioural engine that is capable of both following the rules of the road and avoid them when necessary.

Self-driving vehicles today are in the prototype phase and the idea is becoming just another technology on the computing industry's parts list. By using automotive vision chips that can be used to help vehicles understand the environment around them by detecting pedestrians, traffic lights, collisions, drowsy drivers, and road lane markings. Those tasks initially are more the sort of thing that would help a driver in unusual circumstances rather than take over full time. But they're a significant step in the gradual shift toward

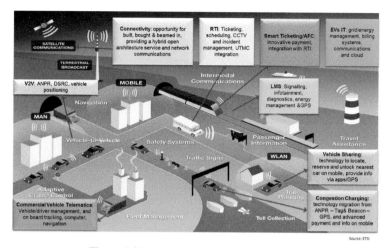

Figure 3.25   ITS Ecosystem (Source: ETSI)

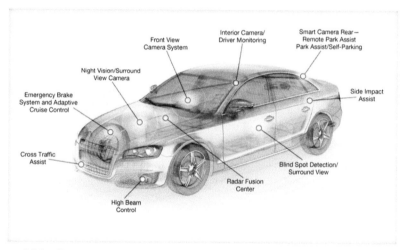

Figure 3.26   Communication and computer vision technologies for driver-assistance and V2V/V2I interaction [80].

the computer-controlled vehicles that Google, Volvo, and other companies are working on [80].

These scenarios are, not independent from each other and show their full potential when combined and used for different applications.

Technical elements of such systems are smart phones and smart vehicle on-board units which acquire information from the user (e.g. position, destination

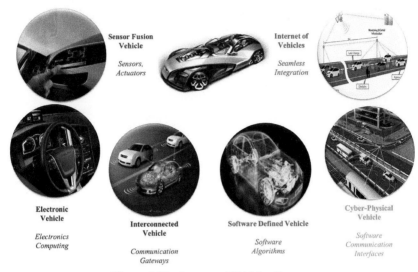

Figure 3.27 Internet of Vehicles Concept

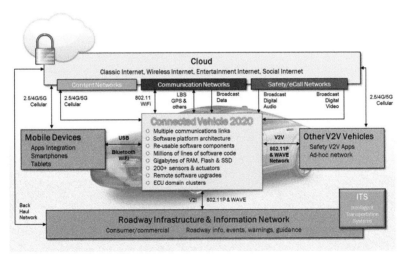

Figure 3.28 Connected Vehicle 2020-Mobility Ecosystem (Source: Continental Corporation)

and schedule) and from on board systems (e.g. vehicle status, position, energy usage profile, driving profile). They interact with external systems (e.g. traffic control systems, parking management, vehicle sharing managements, electric vehicle charging infrastructure). Moreover they need to initiate and perform the related payment procedures.

The concept of Internet of Vehicles (IoV) is the next step for future smart transportation and mobility applications and requires creating new mobile ecosystems based on trust, security and convenience to mobile/contactless services and transportation applications in order to ensure security, mobility and convenience to consumer-centric transactions and services.

Smart sensors in the road and traffic control infrastructures need to collect information about road and traffic status, weather conditions, etc. This requires robust sensors (and actuators) which are able to reliably deliver information to the systems mentioned above. Such reliable communication needs to be based on M2M communication protocols which consider the timing, safety, and security constraints. The expected high amount of data will require sophisticated data mining strategies. Overall optimisation of traffic flow and energy usage may be achieved by collective organisation among the individual vehicles. First steps could be the gradual extension of DATEX-II by IoT related technologies and information. The (international) standardisation of protocol stacks and interfaces is of utmost importance to enable economic competition and guarantee smooth interaction of different vendor products.

When dealing with information related to individuals' positions, destinations, schedules, and user habits, privacy concerns gain highest priority. They even might become road blockers for such technologies. Consequently not only secure communication paths but also procedures which guarantee anonymity and de-personalization of sensible data are of interest.

Some research challenges:

- Safe and secure communication with elements at the network edge, inter-vehicle communication, and vehicle to infrastructure communication
- Energy saving robust and reliable smart sensors and actuators in vehicles and infrastructure
- Technologies for data anonymity addressing privacy concerns
- System partitioning (local/cloud based intelligence)
- Identifying and monitoring critical system elements. Detecting critical overall system states in due time
- Technologies supporting self-organisation and dynamic formation of structures / re-structuring
- Ensure an adequate level of trust and secure exchange of data among different vertical ICT infrastructures (e.g., intermodal scenario).

3.3.4 Smart Home, Smart Buildings and Infrastructure

The rise of Wi-Fi's role in home automation has primarily come about due to the networked nature of deployed electronics where electronic devices (TVs and AV receivers, mobile devices, etc.) have started becoming part of the home IP network and due the increasing rate of adoption of mobile computing devices (smartphones, tablets, etc.).

Several organizations are working to equip homes with technology that enables the occupants to use a single device to control all electronic devices and appliances. The solutions focus primarily on environmental monitoring, energy management, assisted living, comfort, and convenience. The solutions are based on open platforms that employ a network of intelligent sensors to provide information about the state of the home. These sensors monitor systems such as energy generation and metering; heating, ventilation, and air conditioning (HVAC); lighting; security; and environmental key performance indicators. The information is processed and made available through a number of access methods such as touch screens, mobile phones, and 3–D browsers [110]. The networking aspects are bringing online streaming services or network playback, while becoming a mean to control of the device functionality over the network. At the same time mobile devices ensure that consumers have access to a portable 'controller' for the electronics connected to the network. Both types of devices can be used as gateways for IoT applications. In this context many companies are considering building platforms that

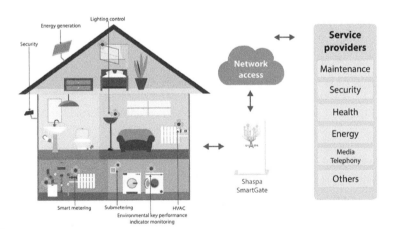

Figure 3.29   Integrated equipment and appliances [109].

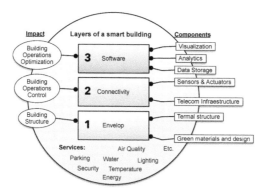

Figure 3.30 Smart Buildings Layers [36]

integrate the building automation with entertainment, healthcare monitoring, energy monitoring and wireless sensor monitoring in the home and building environments.

IoT applications using sensors to collect information about operating conditions combined with cloud hosted analytics software that analyse disparate data points will help facility managers become far more proactive about managing buildings at peak efficiency.

From the technological point of view, it is possible to identify the different layers of a smart building in more detail, to understand the correlation of the systems, services, and management operations. For each layer, is important to understand the implied actors, stakeholders and best practices to implement different technological solutions [36].

Issues of building ownership (i.e., building owner, manager, or occupants) challenge integration with questions such as who pays initial system cost and who collects the benefits over time. A lack of collaboration between the subsectors of the building industry slows new technology adoption and can prevent new buildings from achieving energy, economic and environmental performance targets.

From the layers of a smart building there are many integrated services that can be seen as subsystems. The set of services are managed to provide the best conditions for the activities of the building occupants. The figure below presents the taxonomy of basic services.

Integration of cyber physical systems both within the building and with external entities, such as the electrical grid, will require stakeholder cooperation to achieve true interoperability. As in all sectors, maintaining security will be a critical challenge to overcome [83].

Figure 3.31 Smart Building Services Taxonomy [36]

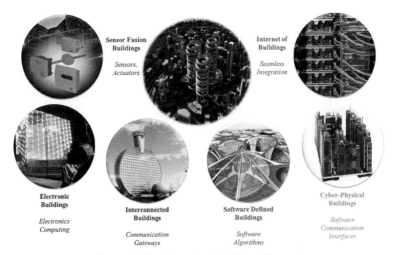

Figure 3.32 Internet of Buildings Concept

Within this field of research the exploitation of the potential of wireless sensor networks (WSNs) to facilitate intelligent energy management in buildings, which increases occupant comfort while reducing energy demand, is highly relevant. In addition to the obvious economic and environmental gains from the introduction of such intelligent energy management in buildings other positive effects will be achieved. Not least of which is the simplification of building control; as placing monitoring, information feedback equipment and control capabilities in a single location will make a buildings' energy management system easier to handle for the building owners, building managers, maintenance crews and other users of the building.

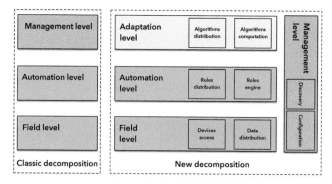

Figure 3.33 Level based architecture of building automation systems [48]

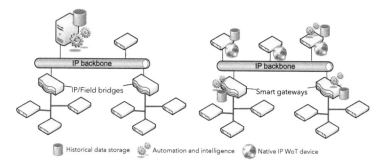

Figure 3.34 Role distribution for a classical building automation system and for a Web-of-Things architecture [48]

Using the Internet together with energy management systems also offers an opportunity to access a buildings' energy information and control systems from a laptop or a Smartphone placed anywhere in the world. This has a huge potential for providing the managers, owners and inhabitants of buildings with energy consumption feedback and the ability to act on that information.

The perceived evolution of building system architectures includes an adaptation level that will dynamically feed the automation level with control logic, i.e. rules. Further, in the IoT approach, the management level has also to be made available transversally as configuration; discovery and monitoring services must be made accessible to all levels. Algorithms and rules have also to be considered as Web resources in a similar way as for sensors and actuators. The repartition of roles for a classical building automation system to the new web of things enabled architecture is different and in this context, future works will have to be carried on to find solutions to minimize the transfer of data and the distribution of algorithms [48].

In the context of the future 'Internet of Things', Intelligent Building Management Systems can be considered part of a much larger information system. This system is used by facilities managers in buildings to manage energy use and energy procurement and to maintain buildings systems. It is based on the infrastructure of the existing Intranets and the Internet, and therefore utilises the same standards as other IT devices. Within this context reductions in the cost and reliability of WSNs are transforming building automation, by making the maintenance of energy efficient, healthy, productive work spaces in buildings increasingly cost effective [72].

3.3.5 Smart Factory and Smart Manufacturing

The role of the Internet of Things is becoming more prominent in enabling access to devices and machines, which in manufacturing systems, were hidden in well-designed silos. This evolution will allow the IT to penetrate further the digitized manufacturing systems. The IoT will connect the factory to a whole new range of applications, which run around the production. This could range from connecting the factory to the smart grid, sharing the production facility as a service or allowing more agility and flexibility within the production systems themselves. In this sense, the production system could be considered one of the many Internets of Things (IoT), where a new ecosystem for smarter and more efficient production could be defined.

The first evolutionary step towards a shared smart factory could be demonstrated by enabling access to today's external stakeholders in order to interact with an IoT-enabled manufacturing system. These stakeholders could include the suppliers of the productions tools (e.g. machines, robots), as well as the production logistics (e.g. material flow, supply chain management), and maintenance and re-tooling actors. An IoT-based architecture that challenges the hierarchical and closed factory automation pyramid, by allowing the above-mentioned stakeholders to run their services in multiple tier flat production system is proposed in [199]. This means that the services and applications of tomorrow do not need to be defined in an intertwined and strictly linked manner to the physical system, but rather run as services in a shared physical world. The room for innovation in the application space could be increased in the same degree of magnitude as this has been the case for embedded applications or Apps, which have exploded since the arrival

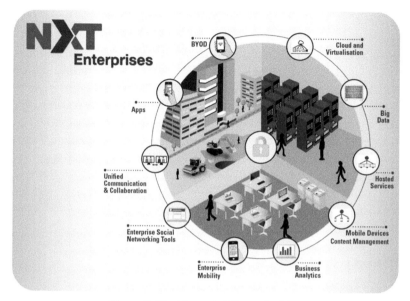

Figure 3.35 Connected Enterprise [61]

of smart phones (i.e. the provision of a clear and well standardized interface to the embedded hardware of a mobile phone to be accessed by all types of Apps).

Enterprises are making use of the huge amount of data available, business analytics, cloud services, enterprise mobility and many others to improve the way businesses are being conducted. These technologies include big data and business analytics software, cloud services, embedded technology, sensor networks / sensing technology, RFID, GPS, M2M, mobility, security and ID recognition technology, wireless network and standardisation.

One key enabler to this ICT-driven smart and agile manufacturing lies in the way we manage and access the physical world, where the sensors, the actuators, and also the production unit should be accessed, and managed in the same or at least similar IoT standard interfaces and technologies. These devices are then providing their services in a well-structured manner, and can be managed and orchestrated for a multitude of applications running in parallel.

The convergence of microelectronics and micromechanical parts within a sensing device, the ubiquity of communications, the rise of micro-robotics, the customization made possible by software will significantly change the world of manufacturing. In addition, broader pervasiveness of telecommunications in many environments is one of the reasons why these environments take the shape of ecosystems.

Some of the main challenges associated with the implementation of cyber-physical systems in include affordability, network integration, and the interoperability of engineering systems.

Most companies have a difficult time justifying risky, expensive, and uncertain investments for smart manufacturing across the company and factory level. Changes to the structure, organization, and culture of manufacturing occur slowly, which hinders technology integration. Pre-digital age control systems are infrequently replaced because they are still serviceable. Retrofitting these existing plants with cyber-physical systems is difficult and expensive. The lack of a standard industry approach to production management results in customized software or use of a manual approach. There is also a need for a unifying theory of non-homogeneous control and communication systems [82].

3.3.6 Smart Health

The market for health monitoring devices is currently characterised by application-specific solutions that are mutually non-interoperable and are made up of diverse architectures. While individual products are designed to cost targets, the long-term goal of achieving lower technology costs across current and future sectors will inevitably be very challenging unless a more coherent approach is used. The IoT can be used in clinical care where hospitalized patients whose physiological status requires close attention can be constantly monitored using IoT-driven, noninvasive monitoring. This requires sensors to collect comprehensive physiological information and uses gateways and the cloud to analyze and store the information and then send the analyzed data wirelessly to caregivers for further analysis and review. These techniques improve the quality of care through constant attention and lower the cost of care by eliminating the need for a caregiver to actively engage in data collection and analysis. In addition the technology can be used for remote monitoring using small, wireless solutions connected through the IoT. These solutions can be used to securely capture patient health data from a variety of sensors, apply complex algorithms to analyze the data and then share it through wireless connectivity with medical professionals who can make appropriate health recommendations.

The links between the many applications in health monitoring are:
- gathering of data from sensors
- support user interfaces and displays

- network connectivity for access to infrastructural services
- low power, robustness, durability, accuracy and reliability.

IoT applications are pushing the development of platforms for implementing ambient assisted living (AAL) systems that will offer services in the areas of assistance to carry out daily activities, health and activity monitoring, enhancing safety and security, getting access to medical and emergency systems, and facilitating rapid health support.

The main objective is to enhance life quality for people who need permanent support or monitoring, to decrease barriers for monitoring important health parameters, to avoid unnecessary healthcare costs and efforts, and to provide the right medical support at the right time.

The IoT plays an important role in healthcare applications, from managing chronic diseases at one end of the spectrum to preventing disease at the other.

Challenges exist in the overall cyber-physical infrastructure (e.g., hardware, connectivity, software development and communications), specialized processes at the intersection of control and sensing, sensor fusion and decision making, security, and the compositionality of cyber-physical systems. Proprietary medical devices in general were not designed for interoperation with other medical devices or computational systems, necessitating advancements in networking and distributed communication within cyber-physical architectures. Interoperability and closed loop systems appears to be the key for success. System security will be critical as communication of individual

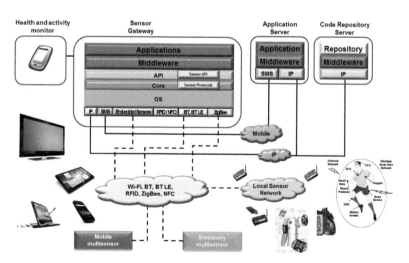

Figure 3.36 Smart Health Platform

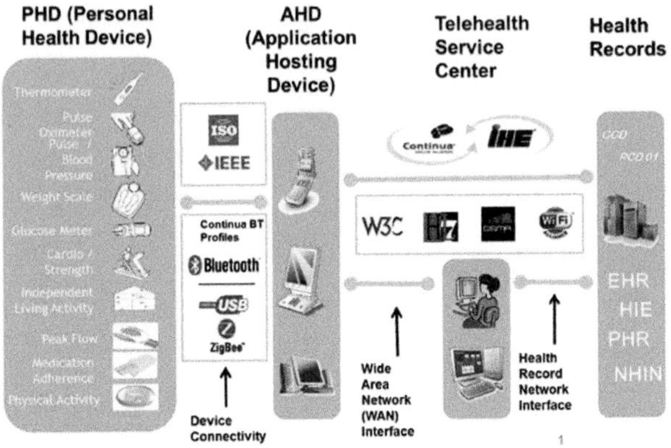

Figure 3.37 Interoperable standard interfaces in the Continua Personal Health Eco-System (Source: Continua Health Alliance)

patient data is communicated over cyber-physical networks. In addition, validating data acquired from patients using new cyber-physical technologies against existing gold standard data acquisition methods will be a challenge. Cyber-physical technologies will also need to be designed to operate with minimal patient training or cooperation [83].

New and innovative technologies are needed to cope with the trends on wired, wireless, high-speed interfaces, miniaturization and modular design approaches for products having multiple technologies integrated.

Internet of Things applications have a future market potential for electronic health services and connected telecommunication industry. In this context, the telecommunications can foster the evolution of ecosystems in different application areas. Medical expenditures are in the range of 10% of the European gross domestic product. The market segment of telemedicine, one of lead markets of the future will have growth rates of more than 19%.

The Continua Health Alliance, an industry consortium promoting telehealth and guaranteeing end-to-end interoperability from sensors to health record databases, has defined in its design guidelines, a dual interface for communication with physiological and residential sensors showing a Personal Area Network (PAN) interface based on Bluetooth Low Energy (BLE) standard and its health device profiles, and a Local Area Network (LAN) interface, based on the Zigbee Health Care application profile. The standards are relatively similar in terms of complexity but BLE, tends to have a longer battery life primarily due to the use of short packet overhead and faster data rates, reduced number of packet exchanges for a short discovery/connect time, and skipped communication events, while Zigbee benefits from a longer range and better reliability with the use of a robust modulation scheme (Direct Sequence Spread Spectrum with orthogonal coding and a mesh-like clustered star networking technology)

Convergence of bio parameter sensing, communication technologies and engineering is turning health care into a new type of information industry. In this context the progress beyond state of the art for IoT applications for healthcare is envisaged as follows:

- Standardisation of interface from sensors and MEMS for an open platform to create a broad and open market for bio-chemical innovators.
- Providing a high degree of automation in the taking and processing of information;
- Real-time data over networks (streaming and regular single measurements) to be available to clinicians anywhere on the web with appropriate software and privileges;
- Data travelling over trusted web.
- Reuse of components over smooth progression between low-cost "home health" devices and higher cost "professional" devices.
- Data needs to be interchangeable between all authorised devices in use within the clinical care pathway, from home, ambulance, clinic, GP, hospital, without manual transfer of data.

3.3.7 Food and Water Tracking and Security

Food and fresh water are the most important natural resources in the world. Organic food produced without addition of certain chemical substances and according to strict rules, or food produced in certain geographical areas will be particularly valued. Similarly, fresh water from mountain springs is already highly valued. In the future it will be very important to bottle and distribute water adequately. This will inevitably lead to attempts to forge the origin or the production process. Using IoT in such scenarios to secure tracking of food or water from the production place to the consumer is one of the important topics.

This has already been introduced to some extent in regard to beef meat. After the "mad cow disease" outbreak in the late 20th century, some beef manufacturers together with large supermarket chains in Ireland are offering "from pasture to plate" traceability of each package of beef meat in an attempt to assure consumers that the meat is safe for consumption. However, this is limited to certain types of food and enables tracing back to the origin of the food only, without information on the production process.

IoT applications need to have a development framework that will assure the following:

- The things connected to the Internet need to provide value. The things that are part of the IoT need to provide a valuable service at a price point that enables adoption, or they need to be part of a larger system that does.
- Use of rich ecosystem for the development. The IoT comprises things, sensors, communication systems, servers, storage, analytics, and end user services. Developers, network operators, hardware manufacturers, and software providers need to come together to make it work. The partnerships among the stakeholders will provide functionality easily available to the customers.
- Systems need to provide APIs that let users take advantage of systems suited to their needs on devices of their choice. APIs also allow developers to innovate and create something interesting using the system's data and services, ultimately driving the system's use and adoption.
- Developers need to be attracted since the implementation will be done on a development platform. Developers using different tools to develop solutions, which work across device platforms playing a key role for future IoT deployment.
- Security needs to be built in. Connecting things previously cut off from the digital world will expose them to new attacks and challenges.

The research challenges are:

- Design of secure, tamper-proof and cost-efficient mechanisms for tracking food and water from production to consumers, enabling immediate notification of actors in case of harmful food and communication of trusted information.
- Secure way of monitoring production processes, providing sufficient information and confidence to consumers. At the same time details of the production processes which might be considered as intellectual property, should not be revealed.
- Ensure trust and secure exchange of data among applications and infrastructures (farm, packing industry, retailers) to prevent the introduction of false or misleading data, which can affect the health of the citizens or create economic damage to the stakeholders.

3.3.8 Participatory Sensing

People live in communities and rely on each other in everyday activities. Recommendations for a good restaurant, car mechanic, movie, phone plan etc. were and still are some of the things where community knowledge helps us in determining our actions.

While in the past this community wisdom was difficult to access and often based on inputs from a handful of people, with the proliferation of the web and more recently social networks, the community knowledge has become readily available - just a click away.

Today, the community wisdom is based on conscious input from people, primarily based on opinions of individuals. With the development of IoT technology and ICT in general, it is becoming interesting to expand the concept of community knowledge to automated observation of events in the real world.

One application of participatory sensing is as a tool for health and wellness, where individuals can self-monitor to observe and adjust their medication, physical activity, nutrition, and interactions. Potential contexts include chronic-disease management and health behaviour change. Communities and health professionals can also use participatory approaches to better understand the development and effective treatment of disease. The same systems can be used as tools for sustainability. Individuals and communities can explore their transportation and consumption habits, and corporations can promote more sustainable practices among employees. In addition, participatory sensing offers a powerful "make a case" technique to support advocacy and civic engagement. It can provide a framework in which citizens can bring to light a civic bottleneck, hazard, personal-safety concern, cultural asset, or other data relevant to urban and natural-resources planning and services, all using data that are systematic and can be validated [121].

Smart phones are already equipped with a number of sensors and actuators: camera, microphone, accelerometers, temperature gauge, speakers, displays etc. A range of other portable sensing products that people will carry in their pockets will soon become available as well. Furthermore, our cars are equipped with a range of sensors capturing information about the car itself, and also about the road and traffic conditions.

Intel is working to simplify deployment of the Internet of Things (IoT) with its Intelligent Systems Framework (Intel®ISF), a set of interoperable solutions designed to address connecting, managing, and securing devices and data in a consistent and scalable manner.

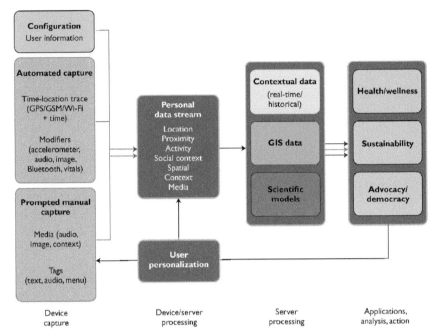

Figure 3.38 Common architectural components for participatory-sensing applications, including mobile device data capture, personal data stream storage, and leveraged data processing [121]

Participatory sensing applications aim at utilizing each person, mobile phone, and car and associated sensors as automatic sensory stations taking a multi-sensor snapshot of the immediate environment. By combining these individual snapshots in an intelligent manner it is possible to create a clear picture of the physical world that can be shared and for example used as an input to the smart city services decision processes.

However, participatory sensing applications come with a number of challenges that need to be solved:

- Design of algorithms for normalization of observations taking into account the conditions under which the observations were taken. For example temperature measurements will be different if taken by a mobile phone in a pocket or a mobile phone lying on a table;
- Design of robust mechanisms for analysis and processing of collected observations in real time (complex event processing) and generation of

"community wisdom" that can be reliably used as an input to decision taking;

- Reliability and trustworthiness of observed data, i.e. design of mechanisms that will ensure that observations were not tampered with and/or detection of such unreliable measurements and consequent exclusion from further processing. In this context, the proper identification and authentication of the data sources is an important function;
- Ensuring privacy of individuals providing observations
- Efficient mechanisms for sharing and distribution of "community wisdom".
- Addressing scalability and large scale deployments

3.3.9 Smart Logistics and Retail

The Internet of Things creates opportunities to achieve efficient solutions in the retail sector by addressing the right person, right content at the right time and right place.

A personalized connected experience is what users are looking for in today's digital environment. Connectivity is key to be connected anytime, anywhere with any devices.

Adapting to the tastes and priorities of changing populations will be a critical task for retailers worldwide.

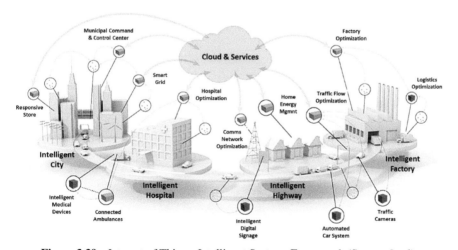

Figure 3.39 Internet of Things: Intelligent Systems Framework (Source: Intel)

To keep up with all these changes, retailers must deploy smart, connected devices throughout their operations.

By tying together everything from inventory tracking to advertising, retailers can gain visibility into their operations and nimbly respond to shifts in consumer behaviour. The challenge is finding a scalable, secure, manageable path to deploying all of these systems.

Retailers are also using sensors, beacons, scanning devices, and other IoT technologies to optimize internally: inventory, fleet, resource, and partner management through real-time analytics, automatic replenishment, notifications, store layout, and more. The Big data generated now affords retailers a factual understanding of how their products, customers, affiliates, employees, and external factors come together. Altogether, this is a $1.6T opportunity for retailers, with $81B in value already realized in 2013 [64].

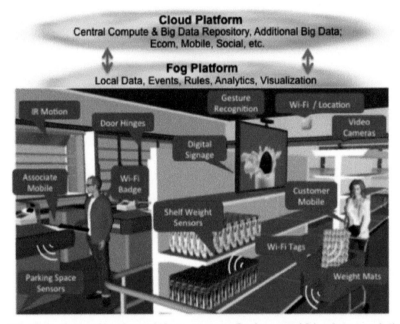

Flexible, hyper-local, real-time, sensor fusion, and big data analytics driving the next generation of Retail Value Chains

Figure 3.40   The Digital Retail Store (Source: Cisco)

3.4 Internet of Things and Related Future Internet Technologies

3.4.1 Cloud Computing

Since the publication of the 2011 SRA, cloud computing has been established as one of the major building blocks of the Future Internet. New technology enablers have progressively fostered virtualisation at different levels and have allowed the various paradigms known as "Applications as a Service", "Platforms as a Service" and "Infrastructure and Networks as a Service". Such trends have greatly helped to reduce cost of ownership and management of associated virtualised resources, lowering the market entry threshold to new players and enabling provisioning of new services. With the virtualisation of objects being the next natural step in this trend, the convergence of cloud computing and Internet of Things will enable unprecedented opportunities in the IoT services arena [104].

As part of this convergence, IoT applications (such as sensor-based services) will be delivered on-demand through a cloud environment [105]. This extends beyond the need to virtualize sensor data stores in a scalable fashion. It asks for virtualization of Internet-connected objects and their ability to become orchestrated into on-demand services (such as Sensing-as-a-Service).

Inadequate security will be a critical barrier to large-scale deployment of IoT systems and broad customer adoption of IoT applications. Simply extending existing IT security architectures to the IoT will not be sufficient. The connected things in the future will have limited resources that can't be easily or cost-effectively upgraded. In order to protect these things over a very long lifespan, this increases the importance of cloud-based security services with resource-efficient, thing-to-cloud interactions. With the growth of IoT, we're shifting toward a cyber-physical paradigm, where we closely integrate

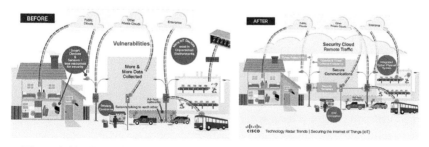

Figure 3.41   Securely Integrating the Cyber and Physical Worlds (Source: Cisco)

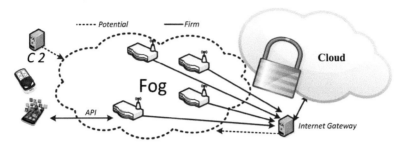

Figure 3.42   Fog Computing Paradigm

computing and communication with the connected things, including the ability to control their operations. In such systems, many security vulnerabilities and threats come from the interactions between the cyber and physical domains. An approach to holistically integrate security vulnerability analysis and protections in both domains will become increasingly necessary. There is growing demand to secure the rapidly increasing population of connected, and often mobile, things. In contrast to today's networks, where assets under protection are typically inside firewalls and protected with access control devices, many things in the IoT arena will operate in unprotected or highly vulnerable environments (i.e. vehicles, sensors, and medical devices used in homes and embedded on patients). Protecting such things poses additional challenges beyond enterprise networks [59].

Many Internet of Things applications require mobility support and geo-distribution in addition to location awareness and low latency, while the data need to be processed in "real-time" in micro clouds or fog. Micro cloud or Fog computing enables new applications and services applies a different data management and analytics and extends the Cloud Computing paradigm to the edge of the network. Similar to Cloud, Micro Cloud/Fog provides data, compute, storage, and application services to end-users.

The Micro Cloud or the fog needs to have the following features in order to efficiently implement the required IoT applications:

- Low latency and location awareness;
- Wide-spread geographical distribution;
- Mobility;
- Very large number of nodes,
- Predominant role of wireless access,
- Strong presence of streaming and real time applications,
- Heterogeneity.

Moreover, generalising the serving scope of an Internet-connected object beyond the "sensing service", it is not hard to imagine virtual objects that will be integrated into the fabric of future IoT services and shared and reused in different contexts, projecting an "Object as a Service" paradigm aimed as in other virtualised resource domains at minimising costs of ownership and maintenance of objects, and fostering the creation of innovative IoT services.

Relevant topics for the research agenda will therefore include:

- The description of requests for services to a cloud/IoT infrastructure,
- The virtualization of objects,
- Tools and techniques for optimization of cloud infrastructures subject to utility and SLA criteria,
- The investigation of utility metrics and (reinforcement) learning techniques that could be used for gauging on-demand IoT services in a cloud environment,
- Techniques for real-time interaction of Internet-connected objects within a cloud environment through the implementation of lightweight interactions and the adaptation of real-time operating systems.
- Access control models to ensure the proper access to the data stored in the cloud.

3.4.2 IoT and Semantic Technologies

The previous IERC SRIAs have identified the importance of semantic technologies towards discovering devices, as well as towards achieving semantic interoperability. Future research on IoT is likely to embrace the concept of Linked Open Data. This could build on the earlier integration of ontologies (e.g., sensor ontologies) into IoT infrastructures and applications.

Semantic technologies will also have a key role in enabling sharing and re-use of virtual objects as a service through the cloud, as illustrated in the previous paragraph. The semantic enrichment of virtual object descriptions will realise for IoT what semantic annotation of web pages has enabled in the Semantic Web. Associated semantic-based reasoning will assist IoT users to more independently find the relevant proven virtual objects to improve the performance or the effectiveness of the IoT applications they intend to use.

3.5 Networks and Communication

Present communication technologies span the globe in wireless and wired networks and support global communication by globally-accepted communication standards. The Internet of Things Strategic Research and Innovation Agenda (SRIA) intends to lay the foundations for the Internet of Things to be developed by research through to the end of this decade and for subsequent innovations to be realised even after this research period. Within this timeframe the number of connected devices, their features, their distribution and implied communication requirements will develop; as will the communication infrastructure and the networks being used. Everything will change significantly. Internet of Things devices will be contributing to and strongly driving this development.

Changes will first be embedded in given communication standards and networks and subsequently in the communication and network structures defined by these standards.

3.5.1 Networking Technology

Mobile traffic today is driven by predictable activities such as making calls, receiving email, surfing the web, and watching videos. Over the next 5 to 10 years, billions of IoT devices with less predictable traffic patterns will join the network, including vehicles, machine-to-machine (M2M) modules, video surveillance that requires all the time bandwidth, or different types of sensors sensor that send out tiny bits of data each day. The rise of cloud computing requires new network strategies for fifth evolution of mobile the 5G, which represents clearly a convergence of network access technologies. The architecture of such network has to integrate the needs for IoT applications and to offer seamless integration. To make the IoT and M2M communication possible there is a need for fast, high-capacity networks.

5G networks will deliver 1,000 to 5,000 times more capacity than 3G and 4G networks today and will be made up of cells that support peak rates of between 10 and 100Gbps. They need to be ultra-low latency, meaning it will take data 1–10 milliseconds to get from one designated point to another, compared to 40–60 milliseconds today. Another goal is to separate communications infrastructure and allow mobile users to move seamlessly between 5G, 4G, and WiFi, which will be fully integrated with the cellular network. Networks will also increasingly become programmable, allowing

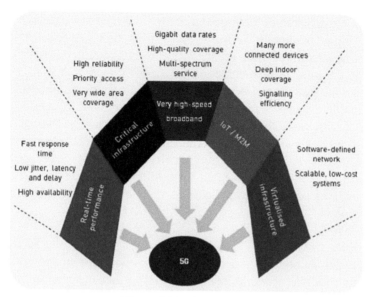

Figure 3.43 5G Features

operators to make changes to the network virtually, without touching the physical infrastructure. These features are important for IoT applications.

The evolution and pervasiveness of present communication technologies has the potential to grow to unprecedented levels in the near future by including the world of things into the developing Internet of Things.

Network users will be humans, machines, things and groups of them.

3.5.1.1 Complexity of the networks of the future
A key research topic will be to understand the complexity of these future networks and the expected growth of complexity due to the growth of Internet of Things. The research results of this topic will give guidelines and timelines for defining the requirements for network functions, for network management, for network growth and network composition and variability [150].

Wireless networks cannot grow without such side effects as interference.

3.5.1.2 Growth of wireless networks
Wireless networks especially will grow largely by adding vast amounts of small Internet of Things devices with minimum hardware, software and intelligence, limiting their resilience to any imperfections in all their functions.

Based on the research of the growing network complexity, caused by the Internet of Things, predictions of traffic and load models will have to guide further research on unfolding the predicted complexity to real networks, their standards and on-going implementations.

Mankind is the maximum user group for the mobile phone system, which is the most prominent distributed system worldwide besides the fixed telephone system and the Internet. Obviously the number of body area networks [36], [151], [152], and of networks integrated into clothes and further personal area networks – all based on Internet of Things devices - will be of the order of the current human population. They are still not unfolding into reality. In a second stage cross network cooperative applications are likely to develop, which are not yet envisioned.

3.5.1.3 Mobile networks

Applications such as body area networks may develop into an autonomous world of small, mobile networks being attached to their bearers and being connected to the Internet by using a common point of contact. The mobile phone of the future could provide this function.

Analysing worldwide industrial processes will be required to find limiting set sizes for the number of machines and all things being implied or used within their range in order to develop an understanding of the evolution steps to the Internet of Things in industrial environments.

3.5.1.4 Expanding current networks to future networks

Generalizing the examples given above, the trend may be to expand current end user network nodes into networks of their own or even a hierarchy of networks. In this way networks will grow on their current access side by unfolding these outermost nodes into even smaller, attached networks, spanning the Internet of Things in the future. In this context networks or even networks of networks will be mobile by themselves.

3.5.1.5 Overlay networks

Even if network construction principles should best be unified for the worldwide Internet of Things and the networks bearing it, there will not be one unified network, but several. In some locations even multiple networks overlaying one another physically and logically.

The Internet and the Internet of Things will have access to large parts of these networks. Further sections may be only represented by a top access node or may not be visible at all globally. Some networks will by intention be shielded against external access and secured against any intrusion on multiple levels.

3.5.1.6 Network self-organization

Wireless networks being built for the Internet of Things will show a large degree of ad-hoc growth, structure, organization, and significant change in time, including mobility. These constituent features will have to be reflected in setting them up and during their operation [153].

Self-organization principles will be applied to configuration by context sensing, especially concerning autonomous negotiation of interference management and possibly cognitive spectrum usage, by optimization of network structure and traffic and load distribution in the network, and in self-healing of networks. All will be done in heterogeneous environments, without interaction by users or operators.

3.5.1.7 IPv6, IoT and Scalability

The current transition of the global Internet to IPv6 will provide a virtually unlimited number of public IP addresses able to provide bidirectional and symmetric (true M2M) access to Billions of smart things. It will pave the way to new models of IoT interconnection and integration. It is raising numerous questions: How can the Internet infrastructure cope with a highly heterogeneous IoT and ease a global IoT interconnection? How interoperability will happen with legacy systems? What will be the impact of the transition to IPv6 on IoT integration, large scale deployment and interoperability? It will probably require developing an IPv6-based European research infrastructure for the IoT.

3.5.1.8 Green networking technology

Network technology has traditionally developed along the line of predictable progress of implementation technologies in all their facets. Given the enormous expected growth of network usage and the number of user nodes in the future, driven by the Internet of Things, there is a real need to minimize the resources for implementing all network elements and the energy being used for their operation [154].

Disruptive developments are to be expected by analysing the energy requirements of current solutions and by going back to principles of communication in wired, optical and wireless information transfer. Research done by Bell Labs [155][156] in recent years shows that networks can achieve an energy efficiency increase of a factor of 1,000 compared to current technologies [157].

The results of the research done by the GreenTouch consortium [155] should be integrated into the development of the network technologies of the future. These network technologies have to be appropriate to realise the Internet of Things and the Future Internet in their most expanded state to be anticipated by the imagination of the experts.

3.5.2 Communication Technology

3.5.2.1 Unfolding the potential of communication technologies

The research aimed at communication technology to be undertaken in the coming decade will have to develop and unfold all potential communication profiles of Internet of Things devices, from bit-level communication to continuous data streams, from sporadic connections to connections being always on, from standard services to emergency modes, from open communication to fully secured communication, spanning applications from local to global, based on single devices to globally-distributed sets of devices [158].

In this context the growth in mobile device market is pushing the deployment of Internet of Things applications where these mobile devices (smart phones, tablets, etc.) are seen as gateways for wireless sensors and actuators.

Based on this research the anticipated bottlenecks in communications and in networks and services will have to be quantified using appropriate theoretical methods and simulation approaches.

Communications technologies for the Future Internet and the Internet of Things will have to avoid such bottlenecks by construction not only for a given status of development, but for the whole path to fully developed and still growing nets.

3.5.2.2 Correctness of construction

Correctness of construction [159] of the whole system is a systematic process that starts from the small systems running on the devices up to network and distributed applications. Methods to prove the correctness of structures and of transformations of structures will be required, including protocols of communication between all levels of communication stacks used in the Internet of Things and the Future Internet.

These methods will be essential for the Internet of Things devices and systems, as the smallest devices will be implemented in hardware and many types will not be programmable. Interoperability within the Internet of Things will be a challenge even if such proof methods are used systematically.

3.5.2.3 An unified theoretical framework for communication

Communication between processes [160] running within an operating system on a single or multicore processor, communication between processes running in a distributed computer system [161], and the communication between devices and structures in the Internet of Things and the Future Internet using wired and wireless channels shall be merged into a unified minimum theoretical framework covering and including formalized communication within protocols.

In this way minimum overhead, optimum use of communication channels and best handling of communication errors should be achievable. Secure communication could be embedded efficiently and naturally as a basic service.

3.5.2.4 Energy-limited Internet of Things devices and their communication

Many types of Internet of Things devices will be connected to the energy grid all the time; on the other hand a significant subset of Internet of Things devices will have to rely on their own limited energy resources or energy harvesting throughout their lifetime.

Given this spread of possible implementations and the expected importance of minimum-energy Internet of Things devices and applications, an important topic of research will have to be the search for minimum energy, minimum computation, slim and lightweight solutions through all layers of Internet of Things communication and applications.

3.5.2.5 Challenge the trend to complexity

The inherent trend to higher complexity of solutions on all levels will be seriously questioned – at least with regard to minimum energy Internet of Things devices and services.

Their communication with the access edges of the Internet of Things network shall be optimized cross domain with their implementation space and it shall be compatible with the correctness of the construction approach.

3.5.2.6 Disruptive approaches

Given these special restrictions, non-standard, but already existing ideas should be carefully checked again and be integrated into existing solutions, and disruptive approaches shall be searched and researched with high priority.

This very special domain of the Internet of Things may well develop into its most challenging and most rewarding domain – from a research point of view and, hopefully, from an economical point of view as well.

3.6 Processes

The deployment of IoT technologies will significantly impact and change the way enterprises do business as well as interactions between different parts of the society, affecting many processes. To be able to reap the many potential benefits that have been postulated for the IoT, several challenges regarding the modelling and execution of such processes need to be solved in order to see wider and in particular commercial deployments of IoT [162]. The special characteristics of IoT services and processes have to be taken into account and it is likely that existing business process modelling and execution languages as well as service description languages such as USDL [165], will need to be extended.

3.6.1 Adaptive and Event-Driven Processes

One of the main benefits of IoT integration is that processes become more adaptive to what is actually happening in the real world. Inherently, this is based on events that are either detected directly or by real-time analysis of sensor data. Such events can occur at any time in the process. For some of the events, the occurrence probability is very low: one knows that they might occur, but not when or if at all. Modelling such events into a process is cumbersome, as they would have to be included into all possible activities, leading to additional complexity and making it more difficult to understand the modelled process, in particular the main flow of the process (the 80% case). Secondly, how to react to a single event can depend on the context, i.e. the set of events that have been detected previously.

Research on adaptive and event-driven processes could consider the extension and exploitation of EDA (Event Driven Architectures) for activity monitoring and complex event processing (CEP) in IoT systems. EDA could be combined with business process execution languages in order to trigger specific steps or parts of a business process.

3.6.2 Processes Dealing with Unreliable Data

When dealing with events coming from the physical world (e.g., via sensors or signal processing algorithms), a degree of unreliability and uncertainty is introduced into the processes. If decisions in a business process are to be taken based on events that have some uncertainty attached, it makes sense to associate each of these events with some value for the quality of information (QoI). In simple cases, this allows the process modeller to define thresholds: e.g., if the degree of certainty is more than 90%, then it is assumed that the event really happened. If it is between 50% and 90%, some other activities will be triggered to determine if the event occurred or not. If it is below 50%, the event is ignored. Things get more complex when multiple events are involved: e.g., one event with 95% certainty, one with 73%, and another with 52%. The underlying services that fire the original events have to be programmed to attach such QoI values to the events. From a BPM perspective, it is essential that such information can be captured, processed and expressed in the modelling notation language, e.g. BPMN. Secondly, the syntax and semantics of such QoI values need to be standardized. Is it a simple certainty percentage as in the examples above, or should it be something more expressive (e.g., a range within which the true value lies)? Relevant techniques should not only address uncertainty in the flow of a given (well-known) IoT-based business process, but also in the overall structuring and modelling of (possibly unknown or unstructured) process flows. Techniques for fuzzy modelling of data and processes could be considered.

3.6.3 Processes dealing with unreliable resources

Not only is the data from resources inherently unreliable, but also the resources providing the data themselves, e.g., due to the failure of the hosting device. Processes relying on such resources need to be able to adapt to such situations. The first issue is to detect such a failure. In the case that a process is calling a resource directly, this detection is trivial. When we're talking about resources that might generate an event at one point in time (e.g., the resource that monitors the temperature condition within the truck and sends an alert if it has become too hot), it is more difficult. Not having received any event can be because of resource failure, but also because there was nothing to report. Likewise, the quality of the generated reports should be regularly audited for correctness. Some monitoring software is needed to detect such problems; it is unclear though if such software should be part of the BPM execution environment or should be a separate component. Among the research challenges is the synchronization of monitoring processes with run-time actuating processes, given that management planes (e.g., monitoring software) tend to operate at different time scales from IoT processes (e.g., automation and control systems in manufacturing).

3.6.4 Highly Distributed Processes

When interaction with real-world objects and devices is required, it can make sense to execute a process in a decentralized fashion. As stated in [165], the decomposition and decentralization of existing business processes increases scalability and performance, allows better decision making and could even lead to new business models and revenue streams through entitlement management of software products deployed on smart items. For example, in environmental monitoring or supply chain tracking applications, no messages need to be sent to the central system as long as everything is within the defined limits. Only if there is a deviation, an alert (event) needs to be generated, which in turn can lead to an adaptation of the overall process. From a business process modelling perspective though, it should be possible to define the process centrally, including the fact that some activities (i.e., the monitoring) will be done remotely. Once the complete process is modelled, it should then be possible to deploy the related services to where they have to be executed, and then run and monitor the complete process.

Relevant research issues include tools and techniques for the synthesis, the verification and the adaptation of distributed processes, in the scope of a volatile environment (i.e. changing contexts, mobility, internet connected objects/devices that join or leave).

3.7 Data Management

Data management is a crucial aspect in the Internet of Things. When considering a world of objects interconnected and constantly exchanging all types of information, the volume of the generated data and the processes involved in the handling of those data become critical.

A long-term opportunity for wireless communications chip makers is the rise of machine-to-machine (M2M) computing, which one of the enabling technologies for Internet of Things. This technology spans a broad range of applications. Worldwide M2M interconnected devices are on a steady upward march that is expected to surge 10-fold to a global total of 12.5 billion devices by 2020. The resulting forecast in M2M traffic shows a similar trajectory, with traffic predicted to grow 24-fold from 2012–2017, representing a CAGR (Compound Annual Growth Rate) of 89% over the same period. Revenue

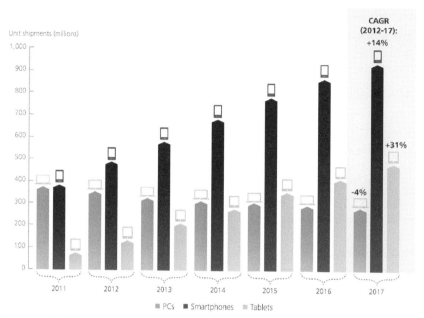

Figure 3.44 PCs, smartphones, and tablets: Unit shipment forecast, worldwide, 2011–2017 [74]

from M2M services spanning a wide range of industry vertical applications, including telematics, health monitoring, smart buildings and security, smart metering, retail point of sale, and retail banking, is set to reach $35 billion by 2016. Driving this surge in the M2M market are a number of forces such as the declining cost of mobile device and infrastructure technology, increased deployment of IP, wireless and wireline networks, and a low-cost opportunity for network carriers to eke out new revenue streams by utilizing existing infrastructure in new markets. This opportunity will likely be most prominent across a number of enterprise verticals, with the energy industry-in the form of smart grid and smart metering technologies-expected to experience significant growth in the M2M market [75].

In this context there are many technologies and factors involved in the "data management" within the IoT context.

Some of the most relevant concepts which enable us to understand the challenges and opportunities of data management are:

- Data Collection and Analysis
- Big data

- Semantic Sensor Networking
- Virtual Sensors
- Complex Event Processing.

3.7.1 Data Collection and Analysis (DCA)

Data Collection and Analysis modules or capabilities are the essential components of any IoT platform or system, and they are constantly evolving in order to support more features and provide more capacity to external components (either higher layer applications leveraging on the data stored by the DCA module or other external systems exchanging information for analysis or processing).

The DCA module is part of the core layer of any IoT platform. Some of the main functions of a DCA module are:

User/customer data storing:
Provides storage of the customer's information collected by sensors

User data & operation modelling:
Allows the customer to create new sensor data models to accommodate collected information and the modelling of the supported operations

On demand data access:
Provides APIs to access the collected data

Device event publish/subscribe/forwarding/notification:
Provides APIs to access the collected data in real time conditions

Customer rules/filtering:
Allows the customer to establish its own filters and rules to correlate events

Customer task automation:
Provides the customer with the ability to manage his automatic processes. (e.g. scheduled platform originated data collection).

Customer workflows:
Allows the customer to create his own workflow to process the incoming events from a device

Multitenant structure:
Provides the structure to support multiple organizations and reseller schemes.

In the coming years, the main research efforts should be targeted to some features that should be included in any Data Collection and Analysis platform:

- **Multi-protocol**. DCA platforms should be capable of handling or understanding different input (and output) protocols and formats. Different standards and wrappings for the submission of observations should be supported

- **De-centralisation**. Sensors and measurements/observations captured by them should be stored in systems that can be de-centralised from a single platform. It is essential that different components, geographically distributed in different locations may cooperate and exchange data. Related with this concept, federation among different systems will make possible the global integration of IoT architectures.
- **Security**. DCA platforms should increase the level of data protection and security, from the transmission of messages from devices (sensors, actuators, etc.) to the data stored in the platform.
- **Data mining** features. Ideally, DCA systems should also integrate capacities for the processing of the stored info, making it easier to extract useful data from the huge amount of contents that may be recorded.

3.7.2 Big Data

Big data is about the processing and analysis of large data repositories, so disproportionately large that it is impossible to treat them with the conventional tools of analytical databases. Some statements suggest that we are entering the "Industrial Revolution of Data," [167], where the majority of data will be stamped out by machines. These machines generate data a lot faster than people can, and their production rates will grow exponentially with Moore's Law. Storing this data is cheap, and it can be mined for valuable information. Examples of this tendency include:

- Web logs;
- RFID;
- Sensor networks;
- Social networks;
- Social data (due to the Social data revolution);
- Internet text and documents;
- Internet search indexing;
- Call detail records;
- Astronomy, atmospheric science, genomics, biogeochemical, biological, and other complex and/or interdisciplinary scientific research;
- Military surveillance;
- Medical records;
- Photography archives;

- Video archives;
- Large scale e-commerce.

The trend is part of an environment quite popular lately: the proliferation of web pages, image and video applications, social networks, mobile devices, apps, sensors, and so on, able to generate, according to IBM, more than 2.5 quintillion bytes per day, to the extent that 90% of the world's data have been created over the past two years.

Big data requires exceptional technologies to efficiently process large quantities of data within a tolerable amount of time. Technologies being applied to big data include massively parallel processing (MPP) databases, data-mining grids, distributed file systems, distributed databases, cloud computing platforms, the Internet, and scalable storage systems. These technologies are linked with many aspects derived from the analysis of natural phenomena such as climate and seismic data to environments such as health, safety or, of course, the business environment.

The biggest challenge of the Petabyte Age will not be storing all that data, it will be figuring out how to make sense of it. Big data deals with unconventional, unstructured databases, which can reach petabytes, exabytes or zettabytes, and require specific treatments for their needs, either in terms of storage or processing/display.

Companies focused on the big data topic, such as Google, Yahoo!, Facebook or some specialised start-ups, currently do not use Oracle tools to process their big data repositories, and they opt instead for an approach based on distributed, cloud and open source systems. An extremely popular example is Hadoop, an Open Source framework in this field that allows applications to work with huge repositories of data and thousands of nodes. These have been inspired by Google tools such as the MapReduce and Google File system,

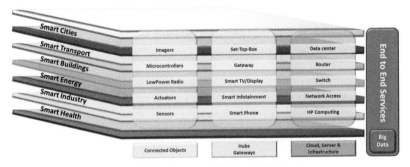

Figure 3.45   Internet of Things holistic view

or NoSQL systems, which in many cases do not comply with the ACID (atomicity, consistency, isolation, durability) characteristics of conventional databases.

In future, it is expected a huge increase in adoption, and many, many questions that must be addressed. Among the imminent research targets in this field are:

- Privacy. Big data systems must avoid any suggestion that users and citizens in general perceive that their privacy is being invaded.
- Integration of both relational and NoSQL systems.
- More efficient indexing, search and processing algorithms, allowing the extraction of results in reduced time and, ideally, near to "real time" scenarios.
- Optimised storage of data. Given the amount of information that the new IoT world may generate, it is essential to avoid that the storage requirements and costs increase exponentially.

3.7.3 Semantic Sensor Networks and Semantic Annotation of data

The information collected from the physical world in combination with the existing resources and services on the Web facilitate enhanced methods to obtain business intelligence, enabling the construction of new types of front-end application and services which could revolutionise the way organisations and people use Internet services and applications in their daily activities. Annotating and interpreting the data, and also the network resources, enables management of the e large scale distributed networks that are often resource and energy constrained, and provides means that allow software agents and intelligent mechanisms to process and reason the acquired data.

There are currently on-going efforts to define ontologies and to create frameworks to apply semantic Web technologies to sensor networks. The Semantic Sensor Web (SSW) proposes annotating sensor data with spatial, temporal, and thematic semantic metadata [169]. This approach uses the current OGC and SWE [171] specifications and attempts to extend them with semantic web technologies to provide enhanced descriptions to facilitate access to sensor data. W3C Semantic Sensor Networks Incubator Group [172] is also working on developing ontology for describing sensors. Effective description of sensor, observation and measurement data and utilising semantic Web technologies for this purpose, are fundamental steps to the construction of semantic sensor networks.

However, associating this data to the existing concepts on the Web and reasoning the data is also an important task to make this information widely available for different applications, front-end services and data consumers.

Semantics allow machines to interpret links and relations between different attributes of a sensor description and also other resources. Utilising and reasoning this information enables the integration of the data as networked knowledge [174]. On a large scale this machine interpretable information (i.e. semantics) is a key enabler and necessity for the semantic sensor networks. Emergence of sensor data as linked-data enables sensor network providers and data consumers to connect sensor descriptions to potentially endless data existing on the Web. By relating sensor data attributes such as location, type, observation and measurement features to other resources on the Web of data, users will be able to integrate physical world data and the logical world data to draw conclusions, create business intelligence, enable smart environments, and support automated decision making systems among many other applications.

The linked-sensor-data can also be queried, accessed and reasoned based on the same principles that apply to linked-data. The principles of using linked data to describe sensor network resources and data in an implementation of an open platform to publish and consume interoperable sensor data is described in [175].

In general, associating sensor and sensor network data with other concepts (on the Web) and reasoning makes the data information widely available for different applications, front-end services and data consumers. The semantic description allow machines to interpret links and relations between the different attributes of a sensor description and also other data existing on the Web or provided by other applications and resources. Utilising and reasoning this information enables the integration of the data on a wider scale, known as networked knowledge [174]. This machine-interpretable information (i.e. semantics) is a key enabler for the semantic sensor networks.

3.7.4 Virtual Sensors

A virtual sensor can be considered as a product of spatial, temporal and/or thematic transformation of raw or other virtual sensor producing data with necessary provenance information attached to this transformation. Virtual sensors and actuators are a programming abstraction simplifying the development of decentralized WSN applications [176].

Models for interacting with wireless sensors such as Internet of Things and sensor cloud aim to overcome restricted resources and efficiency. New sensor clouds need to enable different networks, cover a large geographical area, connect together and be used simultaneously by multiple users on demand. Virtual sensors, as the core of the sensor cloud architecture, assist in creating a multiuser environment on top of resource-constrained physical wireless sensors and can help in supporting multiple applications.

The data acquired by a set of sensors can be collected, processed according to an application-provided aggregation function, and then perceived as the reading of a single virtual sensor. Dually, a virtual actuator provides a single entry point for distributing commands to a set of real actuator nodes. We follow that statement with this definition:

- A virtual sensor behaves just like a real sensor, emitting time-series data from a specified geographic region with newly defined thematic concepts or observations which the real sensors may not have.
- A virtual sensor may not have any real sensor's physical properties such as manufacturer or battery power information, but does have other properties, such as: who created it; what methods are used, and what original sensors it is based on.

3.8 Security, Privacy & Trust

The Internet of Things presents security-related challenges that are identified in the IERC 2010 Strategic Research and Innovation Roadmap but some elaboration is useful as there are further aspects that need to be addressed by the research community. While there are a number of specific security, privacy and trust challenges in the IoT, they all share a number of transverse non-functional requirements:

- Lightweight and symmetric solutions, Support for resource constrained devices
- Scalable to billions of devices/transactions

Solutions will need to address federation/administrative co-operation

- Heterogeneity and multiplicity of devices and platforms
- Intuitively usable solutions, seamlessly integrated into the real world

3.8.1 Trust for IoT

As IoT-scale applications and services will scale over multiple administrative domains and involve multiple ownership regimes, there is a need for a trust framework to enable the users of the system to have confidence that the information and services being exchanged can indeed be relied upon. The trust framework needs to be able to deal with humans and machines as users, i.e. it needs to convey trust to humans and needs to be robust enough to be used by machines without denial of service. The development of trust frameworks that address this requirement will require advances in areas such as:

- Lightweight Public Key Infrastructures (PKI) as a basis for trust management. Advances are expected in hierarchical and cross certification concepts to enable solutions to address the scalability requirements.
- Lightweight key management systems to enable trust relationships to be established and the distribution of encryption materials using minimum communications and processing resources, as is consistent with the resource constrained nature of many IoT devices.
- Quality of Information is a requirement for many IoT-based systems where metadata can be used to provide an assessment of the reliability of IoT data.
- Decentralised and self-configuring systems as alternatives to PKI for establishing trust e.g. identity federation, peer to peer.
- Novel methods for assessing trust in people, devices and data, beyond reputation systems. One example is Trust Negotiation. Trust Negotiation is a mechanism that allows two parties to automatically negotiate, on the basis of a chain of trust policies, the minimum level of trust required to grant access to a service or to a piece of information.
- Assurance methods for trusted platforms including hardware, software, protocols, etc.
- Access Control to prevent data breaches. One example is Usage Control, which is the process of ensuring the correct usage of certain information according to a predefined policy after the access to information is granted.

3.8.2 Security for IoT

As the IoT becomes a key element of the Future Internet and a critical national/international infrastructure, the need to provide adequate security for the IoT infrastructure becomes ever more important.

IoT applications use sensors and actuators embedded in the environment and they collect large volumes of data on room temperatures, humidity, and lighting to optimize energy consumption and avoid operational failures that have a real impact on the environment. In the retail industry, a refrigerator failing to maintain proper cooling temperatures could place high value medical or food inventory at risk. Having all of these devices connected, it is as well needed have the right data model. The data model has to accommodate high data rate sensor data and to assimilate and analyze the information. In this context database read/write performance is critical, particularly with high data rate sensor data. The database must support high-speed read and writes, be continuously available (100% of the time) to gather this data at uniform intervals and be scalable in order to maintain a cost-effective horizontal data store over time.

Large-scale applications and services based on the IoT are increasingly vulnerable to disruption from attack or information theft. Advances are required in several areas to make the IoT secure from those with malicious intent, including

- DoS/DDOS attacks are already well understood for the current Internet, but the IoT is also susceptible to such attacks and will require specific techniques and mechanisms to ensure that transport, energy, city infrastructures cannot be disabled or subverted.
- General attack detection and recovery/resilience to cope with IoT-specific threats, such as compromised nodes, malicious code hacking attacks.
- Cyber situation awareness tools/techniques will need to be developed to enable IoT-based infrastructures to be monitored. Advances are required to enable operators to adapt the protection of the IoT during the lifecycle of the system and assist operators to take the most appropriate protective action during attacks.
- The IoT requires a variety of access control and associated accounting schemes to support the various authorisation and usage models that are required by users. The heterogeneity and diversity of the devices/gateways that require access control will require new lightweight schemes to be developed.
- The IoT needs to handle virtually all modes of operation by itself without relying on human control. New techniques and approaches e.g. from machine learning, are required to lead to a self-managed IoT.

3.8.3 Privacy for IoT

As much of the information in an IoT system may be personal data, there is a requirement to support anonymity and restrictive handling of personal information.

There are a number of areas where advances are required:

- Cryptographic techniques that enable protected data to be stored processed and shared, without the information content being accessible to other parties. Technologies such as homomorphic and searchable encryption are potential candidates for developing such approaches.
- Techniques to support Privacy by Design concepts, including data minimisation, identification, authentication and anonymity.
- Fine-grain and self-configuring access control mechanism emulating the real world

There are a number of privacy implications arising from the ubiquity and pervasiveness of IoT devices where further research is required, including

- Preserving location privacy, where location can be inferred from things associated with people.
- Prevention of personal information inference, that individuals would wish to keep private, through the observation of IoT-related exchanges.
- Keeping information as local as possible using decentralised computing and key management.
- Use of soft Identities, where the real identity of the user can be used to generate various soft identities for specific applications. Each soft identity can be designed for a specific context or application without revealing unnecessary information, which can lead to privacy breaches.

3.9 Device Level Energy Issues

One of the essential challenges in IoT is how to interconnect "things" in an interoperable way while taking into account the energy constraints, knowing that the communication is the most energy consuming task on devices. RF solutions for a wide field of applications in the Internet of Things have been released over the last decade, led by a need for integration and low power consumption.

3.9.1 Low Power Communication

Several low power communication technologies have been proposed from different standardisation bodies. The most common ones are:

- **IEEE 802.15.4** has developed a low-cost, low-power consumption, low complexity, low to medium range communication standard at the link and the physical layers [181] for resource constrained devices.

- **Bluetooth low energy** (Bluetooth LE, [182]) is the ultra-low power version of the Bluetooth technology [183] that is up to 15 times more efficient than Bluetooth.

- **Ultra-Wide Bandwidth (UWB) Technology** [183] is an emerging technology in the IoT domain that transmits signals across a much larger frequency range than conventional systems. UWB, in addition to its communication capabilities, it can allow for high precision ranging of devices in IoT applications.

- **ISO 18000–7 DASH7** standard developed by DASH7 Alliance is a low power, low complexity, radio protocol for all sub 1GHz radio devices. It is a non-proprietary technology based on an open standard, and the solutions may contain a pool of companion technologies operating in their own ways. Common for these technologies are that they use a Sub 1 GHz silicon radio (433 MHz) as their primary communicating device [25]. The applications using DASH7 include supply chain management, inventory/yard management, manufacturing and warehouse optimization, hazardous material monitoring, smart meter and commercial green building development.

- **RFID/NFC** proposes a variety of standards to offer contactless solutions. Proximity cards can only be read from less than 10 cm and follows the ISO 14443 standard [185] and is also the basis of the NFC standard. RFID tags or vicinity tags dedicated to identification of objects have a reading distance which can reach 7 to 8 meters.

Nevertheless, front-end architectures have remained traditional and there is now a demand for innovation. Regarding the ultra-low consumption target, super-regenerative have proven to be very energetically efficient architectures used for Wake-Up receivers. It remains active permanently at very low power consumption, and can trigger a signal to wake up a complete/standard receiver [186–187]. In this field, standardization is required, as today only proprietary solutions exist, for an actual gain in the overall market to be significant.

On the other hand, power consumption reduction of an RF full-receiver can be envisioned, with a target well below 5mW to enable very small form factor and long life-time battery. Indeed, targeting below 1mW would then enable support from energy harvesting systems enabling energy autonomous RF communications. In addition to this improvement, lighter communication protocols should also be envisioned as the frequent synchronization requirement makes frequent activation of the RF link mandatory, thereby overhead in the power consumption.

It must also be considered that recent advances in the area of CMOS technology beyond 90 nm, even 65 nm nodes, leads to new paradigms in the field of RF communication. Applications which require RF connectivity are growing as fast as the Internet of Things, and it is now economically viable to propose this connectivity solution as a feature of a wider solution. It is already the case for the micro-controller which can now easily embed a ZigBee or Bluetooth RF link, and this will expand to meet other large volume applications sensors.

Progressively, portable RF architectures are making it easy to add the RF feature to existing devices. This will lead to RF heavily exploiting digital blocks and limiting analogue ones, like passive / inductor silicon consuming elements, as these are rarely easy to port from one technology to another. Nevertheless, the same performance will be required so receiver architectures will have to efficiently digitalize the signal in the receiver or transmitter chain [188]. In this direction, Band-Pass Sampling solutions are promising as the signal is quantized at a much lower frequency than the Nyquist one, related to deep under-sampling ratio [189]. Consumption is therefore greatly reduced compared to more traditional early-stage sampling processes, where the sampling frequency is much lower.

Continuous-Time quantization has also been regarded as a solution for high-integration and easy portability. It is an early-stage quantization as well, but without sampling [190]. Therefore, there is no added consumption due to the clock, only a signal level which is considered. These two solutions are clear evolutions to pave the way to further digital and portable RF solutions.

Cable-powered devices are not expected to be a viable option for IoT devices as they are difficult and costly to deploy. Battery replacements in devices are either impractical or very costly in many IoT deployment scenarios. As a consequence, for large scale and autonomous IoT, alternative energy sourcing using ambient energy should be considered.

3.9.2 Energy Harvesting

Four main ambient energy sources are present in our environment: mechanical energy, thermal energy, radiant energy and chemical energy. The power consumption varies depending on the communication protocols and data rate used to transmit the date. The approximate power consumption for different protocols is as following 3G-384kbps-2W, GPRS-24kbps-1W, WiFi-10Mbps-32–200mW, Bluetooth-1Mbps-2.5–100 mW, and Zigbee-250kbps-1mW.

Ambient light, thermal gradients, vibration/motion or electromagnetic radiation can be harvested to power electronic devices. The major components of an autonomous wireless sensor are the energy harvesting transducer, energy processing, sensor, microcontroller and the wireless radio. For successful energy harvesting implementations there are three key areas in the energy processing stage that must be addressed: energy conversion, energy storage, and power management.

Harvesting 100 µW during 1 year corresponds to a total amount of energy equivalent to 1 g of lithium. Considering this approach of looking at energy consumption for one measurement instead of average power consumption, it results that, today:

- Sending 100 bits of data consumes about 5 µJ,
- Measuring acceleration consumes about 50 µJ,
- Making a complete measurement: measure + conversion + emission consume 250–500 µJ.

Therefore, with 100 µW harvested continuously, it is possible to perform a complete measurement every 1–10 seconds. This duty cycle can be sufficient for many applications. For other applications, basic functions' power consumptions are expected to be reduced by 10 to 100 within 10 years; which will enable continuous running mode of EH-powered IoT devices.

Even though many developments have been performed over the last 10 years, energy harvesting – except PV cells – is still an emerging technology that has not yet been adopted by industry. Nevertheless, further improvements of present technologies should enable the needs of IoT to be met.

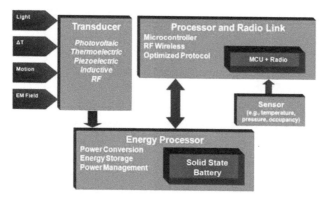

Figure 3.46 Energy harvesting - components of an autonomous wireless sensor (Source: Cymbet)

An example of interoperable wireless standard that enables switches, gateways and sensors from different manufacturers to combine seamlessly and wireless communicates with all major wired bus systems such as KNX, LON, BACnet or TCP/IP is presented in [120].

The development of energy harvesting and storage devices is instrumental to the realization of the ubiquitous connectivity that the IoT proclaims and the potential market for portable energy storage and energy harvesting could be in distributed smart swarms of mobile systems for the Internet of Things.

The energy harvesting wireless sensor solution is able to generate a signal from an extremely small amount of energy. From just 50 μWs a standard energy harvesting wireless module can easily transmit a signal 300 meters (in a free field).

3.9.3 Future Trends and Recommendations

In the future, the number and types of IoT devices will increase, therefore interoperability between devices will be essential. More computation and yet less power and lower cost requirements will have to be met. Technology integration will be an enabler along with the development of even lower power technology and improvement of battery efficiency. The power consumption of computers over the last 60 years was analysed in [192] and the authors concluded that electrical efficiency of computation has doubled roughly every year and a half. A similar trend can be expected for embedded computing using similar technology over the next 10 years. This would lead to a reduction by an order of 100 in power consumption at same level of computation. Allowing for a 10 fold increase in IoT computation, power consumption should still be reduced by an order of 10.

On the other hand, energy harvesting techniques have been explored to respond to the energy consumption requirements of the IoT domain. For vibration energy harvesters, we expect them to have higher power densities in the future (from 10 μW/g to 30 μW/g) and to work on a wider frequency bandwidth. Actually, the goal of vibration energy harvesters' researchers is to develop Plug and Play (PnP) devices, able to work in any vibrating environment, within 10 years. In the same time, we expect basic functions' energy consumption to decrease by at least a factor of 10. All these progresses will allow vibration energy harvesters to attract new markets, from industry to healthcare or defence.

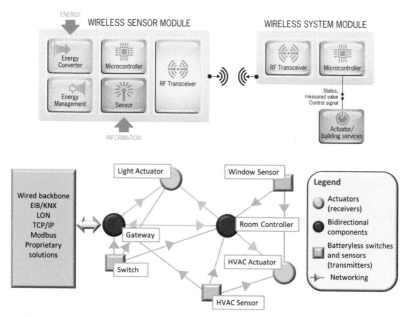

Figure 3.47 Energy harvesting wireless sensor network (Source: EnOcean)

The main challenge for thermoelectric solutions is to increase thermo-electric materials' intrinsic efficiency, in order to convert a higher part of the few mW of thermal energy available. This efficiency improvement will be mainly performed by using micro and nanotechnologies (such as superlattices or quantum dots).

For solar energy harvesting, photovoltaic cells are probably the most advanced and robust solution. They are already used in many applications and for most of them, today's solutions are sufficient. Yet, for IoT devices, it could be interesting to improve the photovoltaic cells efficiency to decrease photovoltaic cells' sizes and to harvest energy in even darker places.

In the future batteries will recharge from radio signals, cell phones will recharge from Wi-Fi. Smaller Cells (micro, pico, femto) will result in more cell sites with less distance apart but they will be greener, provide power/cost savings and at the same time, higher throughput. Connected homes will enable consumers to manage their energy, media, security and appliances; will be part of the IoT applications in the future.

3.10 IoT Related Standardization

The IERC previous SRAs [68] [85] addresses the topic of standardization and is focused on the actual needs of producing specific standards. This chapter examines further standardization considerations.

3.10.1 The Role of Standardization Activities

Standards are needed for interoperability both within and between domains. Within a domain, standards can provide cost efficient realizations of solutions, and a domain here can mean even a specific organization or enterprise realizing an IoT. Between domains, the interoperability ensures cooperation between the engaged domains, and is more oriented towards a proper "Internet of Things". There is a need to consider the life-cycle process in which standardization is one activity. Significant attention is given to the "pre-selection" of standards through collaborative research, but focus should also be given to regulation, legislation, interoperability and certification as other activities in the same life-cycle. For IoT, this is of particular importance.

A complexity with IoT comes from the fact that IoT intends to support a number of different applications covering a wide array of disciplines that are not part of the ICT domain. Requirements in these different disciplines can often come from legislation or regulatory activities. As a result, such policy making can have a direct requirement for supporting IoT standards to be developed. It would therefore be beneficial to develop a wider approach to standardization and include anticipation of emerging or on-going policy making in target application areas, and thus be prepared for its potential impact on IoT-related standardization.

A typical example is the standardization of vehicle emergency call services called eCall driven from the EC [193]. Based on the objective of increased road safety, directives were established that led to the standardization of solutions for services and communication by e.g. ETSI, and subsequently 3GPP. Another example is the Smart Grid standardization mandate M/490 [194] from the EC towards the European Standards Organisations (ESOs), and primarily ETSI, CEN and CENELEC.

The standardization bodies are addressing the issue of interoperable protocol stacks and open standards for the IoT. This includes as well expending the HTTP, TCP, IP stack to the IoT-specific protocol stack. This is quite challenging considering the different wireless protocols like ZigBee, RFID, Bluetooth, BACnet 802.15.4e, 6LoWPAN, RPL, CoAP , AMQP and MQTT.

HTTP relies on the Transmission Control Protocol (TCP). TCP's flow control mechanism is not appropriate for LLNs and its overhead is considered too high for short-lived transactions. In addition, TCP does not have multicast support and is rather sensitive to mobility. CoAP is built on top of the User Datagram Protocol (UDP) and therefore has significantly lower overhead and multicast support [103].

The conclusion is that any IoT related standardization must pay attention to how regulatory measures in a particular applied sector will eventually drive the need for standardized efforts in the IoT domain.

Agreed standards do not necessarily mean that the objective of interoperability is achieved. The mobile communications industry has been successful not only because of its global standards, but also because interoperability can be assured via the certification of mobile devices and organizations such as the Global Certification Forum [195] which is a joint partnership between mobile network operators, mobile handset manufacturers and test equipment manufacturers. Current corresponding M2M efforts are very domain specific and fragmented. The emerging IoT and M2M dependant industries should also benefit from ensuring interoperability of devices via activities such as conformance testing and certification on a broader scale.

To achieve this very important objective of a "certification" or validation programme, we also need non ambiguous test specifications which are also standards. This represents a critical step and an economic issue as this activity is resource consuming. As for any complex technology, implementation of test specifications into cost-effective test tools should also to be considered. A good example is the complete approach of ETSI using a methodology (e.g. based on TTCN-3) considering all the needs for successful certification programmes.

The conclusion therefore is that just as the applied sector can benefit from standards supporting their particular regulated or mandated needs, equally, these sectors can benefit from conforming and certified solutions, protocols and devices. This is certain to help the IoT- supporting industrial players to succeed.

It is worth noting that setting standards for the purpose of interoperability is not only driven by proper SDOs, but for many industries and applied sectors it can also be driven by Special Interest Groups, Alliances and the Open Source communities. It is of equal importance from an IoT perspective to consider these different organizations when addressing the issue of standardization.

From the point of view of standardisation IoT is a global concept, and is based on the idea that anything can be connected at any time from any place to any network, by preserving the security, privacy and safety. The concept of connecting any object to the Internet could be one of the biggest standardization challenges and the success of the IoT is dependent on the development of interoperable global standards. In this context the IERC position is very clear. Global standards are needed to achieve economy of scale and interworking. Wireless sensor networks, RFID, M2M are evolving to intelligent devices which need networking capabilities for a large number of applications and these technologies are "edge" drivers towards the "Internet of Things", while the network identifiable devices will have an impact on telecommunications networks. IERC is focussed to identify the requirements and specifications from industry and the needs of IoT standards in different domains and to harmonize the efforts, avoid the duplication of efforts and identify the standardization areas that need focus in the future.

To achieve these goals it is necessary to overview the international IoT standardization items and associated roadmap; to propose a harmonized European IoT standardisation roadmap; work to provide a global harmonization of IoT standardization activities; and develop a basic framework of standards (e.g., concept, terms, definition, relation with similar technologies).

3.10.2 Current Situation

The current M2M related standards and technologies landscape is highly fragmented. The fragmentation can be seen across different applied domains where there is very little or no re-use of technologies beyond basic communications or networking standards. Even within a particular applied sector, a number of competing standards and technologies are used and promoted. The entire ecosystem of solution providers and users would greatly benefit from less fragmentation and should strive towards the use of a common set of basic tools. This would provide faster time to market, economy of scale and reduce overall costs.

Another view is standards targeting protocols vs. systems. Much emphasis has been put on communications and protocol standards, but very little effort has previously been invested in standardizing system functions or system architectures that support IoT. Localized system standards are plentiful for specific deployments in various domains. One such example is in building automation and control with (competing) standards like BACnet and KNX. However, system standards on the larger deployment and global scale are not in place. The on going work in ETSI M2M TC is one such approach, but is currently limited to providing basic application enablement on top of different networks. It should also be noted that ETSI represent one industry – the telecommunications industry. The IoT stakeholders are

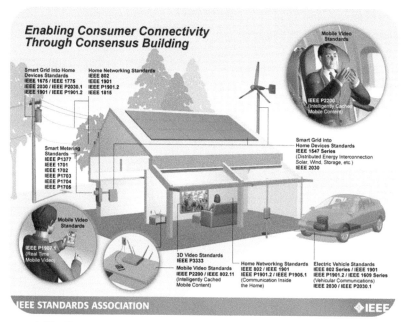

Figure 3.48 Enabling Consumer Connectivity Through Consensus Building (Source: IEEE-SA)

represented by a number of different industries and sectors reaching far beyond telecommunications.

IEEE-SA is also collaborating with other Standards Development Organizations to create a more efficient and collaborative standards-development environment.

Developing smart grids around the world will produce benefits - from the ability to respond to demand with more or less generation, to identifying waste and reducing costs. But it's connecting to what's in the home that will produce the greatest efficiencies, because the homes/buildings are where the grid connects to the user. By bringing the user online, the smart grid can manage demand, eliminate waste, lower peak loads, and stimulate investment in more energy efficient appliances. Utilities, manufacturers and suppliers are using IEEE standards to make the Smart Grid work with their products and the customers' homes/buildings. The standards addressing this area are as following [67]:

- Smart Grid Interoperability — IEEE 2030^{TM}
- Smart Metering — IEEE P1377^{TM}, IEEE 1701^{TM}, IEEE 1702^{TM}, IEEE P1703^{TM}, IEEE P1704^{TM}, IEEE P1705^{TM}

- Utility Network Protocol — IEEE 1815^{TM}
- Interconnecting Distributed Resources with Electrical Power Systems - IEEE 1547^{TM} series
- Communication over Power Lines — IEEE 1901^{TM}, IEEE P1901.2^{TM}
- Local and Metropolitan Area Networks — IEEE 802§series

The electric vehicle will interface with the homes/buildings and the electrical grid is being shaped by the feedback of owners and manufacturers today. The standards addressing this area are as following [67]:

- Smart Grid Interoperability – IEEE 2030^{TM}, IEEE P2030.1^{TM}
- Communication over Power Lines – IEEE 1901^{TM}, IEEE P1901.2^{TM}
- Local and Metropolitan Area Networks – IEEE 802§series
- Interconnecting Distributed Resources with Electrical Power Systems - IEEE 1547^{TM} series
- Smart Metering/Utility Network Protocol – IEEE 1701^{TM}, IEEE 1702^{TM}, IEEE P1703^{TM}, IEEE P1704^{TM}, IEEE P1705^{TM}, IEEE P1377^{TM}, IEEE 1815^{TM}

The IoT will bring home/building networking for connecting devices and humans to communicate. This will empower the devices themselves and allow them to interact. In order to make home/building-wide systems with components from many manufacturers work requires connectivity standards and an assurance of interoperability. The standards addressing this area are as following [67]:

- Convergent Digital Home Network – IEEE P1905.1^{TM}
- Power Lines Communications – IEEE 1901^{TM}, IEEE P1901.2^{TM}, IEEE 1675^{TM}, IEEE 1775^{TM}
- Low-Frequency and Wireless Protocol – IEEE 1902.1^{TM}
- Local and Metropolitan Area Networks – IEEE 802®series
- Utility Network Protocol – IEEE 1815^{TM}

3.10.3 Areas for Additional Consideration

The technology fragmentation mentioned above is particularly evident on the IoT device side. To drive further standardization of device technologies in the direction of standard Internet protocols and Web technologies, and towards the application level, would mitigate the impacts of fragmentation and strive towards true interoperability. Embedded web services, as driven by the IETF and IPSO Alliance, will ensure a seamless integration of IoT devices with the

Internet. It will also need to include semantic representation of IoT device hosted services and capabilities.

The service layer infrastructure will require standardization of necessary capabilities like interfaces to information and sensor data repositories, discovery and directory services and other mechanisms that have already been identified in projects like SENSEI [195], IoT-A [196], and IoT6. Current efforts in ETSI M2M TC do not address these aspects.

The IoT will require federated environments where producers and consumers of services and information can collaborate across both administrative and application domains. This will require standardized interfaces on discovery capabilities as well as the appropriate semantic annotation to ensure that information becomes interoperable across sectors. Furthermore, mechanisms for authentication and authorization as well as provenance of information, ownership and "market mechanisms" for information become particularly important in a federated environment. Appropriate SLAs will be required for standardization. F-ONS [199] is one example activity in the direction of federation by GS1. Similar approaches will be needed in general for IoT including standardized cross-domain interfaces of sensor based services.

A number of IoT applications will be coming from the public sector. The Directive on Public Sector Information [201] requires open access to data. Integration of data coming from various application domains is not an easy task as data and information does not adhere to any standardized formats including their semantics. Even within a single domain, data and information is not easily integrated or shared. Consideration of IoT data and information integration and sharing within domains as well as between domains need, also be considered at the international level.

Instrumental in a number of IoT applications is the spatial dimension. Standardization efforts that provide necessary harmonization and interoperability with spatial information services like INSPIRE [202] will be the key.

IoT with its envisioned billions of devices producing information of very different characteristics will place additional requirements on the underlying communications and networking strata. Efforts are needed to ensure that the networks can accommodate not only the number of devices but also the very different traffic requirements including delay tolerance, latency and reliability. This is of particular importance for wireless access networks which traditionally have been optimized based on a different set of characteristics. 3GPP, as an example, has acknowledged this and has started to address the short term needs, but the long term needs still require identification and standardization.

3.10.4 Interoperability in the Internet-of-Things

The Internet of Things (IoT) is shaping the evolution of the future Internet. After connecting people anytime and everywhere, the next step is to interconnect heterogeneous things / machines / smart objects both between themselves and with the Internet; allowing by thy way, the creation of value-added open and interoperable services/applications, enabled by their interconnection, in such a way that they can be integrated with current and new business and development processes.

As for the IoT, future networks will continue to be heterogeneous, multi-vendors, multi-services and largely distributed. Consequently, the risk of non-interoperability will increase. This may lead to unavailability of some services for end-users that can have catastrophic consequences regarding applications related for instance to emergency or health, etc. Or, it could also mean that users/applications are likely to loose key information out of the IoT due to this lack of interoperability. Thus, it is vital to guarantee that network components will interoperate to unleash the full value of the Internet of Things.

3.10.4.1 IoT Interoperability necessary framework

Interoperability is a key challenge in the realms of the Internet of Things (IoT)! This is due to the intrinsic fabric of the IoT as: (i) high–dimensional, with the co-existence of many systems (devices, sensors, equipment, etc.) in the environment that need to communicate and exchange information; (ii) highly-heterogeneous, where these vast systems are conceived by a lot of manufacturers and are designed for much different purposes and targeting diverse application domains, making it extremely difficult (if not impossible) to reach out for global agreements and widely accepted specification; (iii) dynamic and non-linear, where new Things (that were not even considered at start) are entering (and leaving) the environment all the time and that support new unforeseen formats and protocols but that need to communicate and share data in the IoT; and (iv) hard to describe/model due to existence of many data formats, described in much different languages, that can share (or not) the same modelling principles, and that can be interrelated in many ways with one another. **This qualifies interoperability in the IoT as a problem of complex nature!**

Also, the Internet of Things can be seen as both the first and the final frontier of interoperability. First, as it is the initial mile of a sensing system and where interoperability would enable Things to talk and collaborate altogether for an higher purpose; and final, as it is possibly the place where interoperability is more difficult to tackle due to the unavoidable complexities of the IoT. We therefore need some novel approaches and comprehensions of Interoperability for the Internet of Things also making sure that it endures, that it is sustainable. **It is then needed sustainable interoperability in the Internet of Things!**

This means that we need to cope at the same time with the complex nature and sustainability requirement of interoperability in the Internet of Things. For this, it is needed a framework for sustainable interoperability that especially targets the Internet of Things taking on its specifics and constraints. This framework can (and should) learn from the best-of-breed interoperability solutions from related domains (e.g. enterprise interoperability), to take the good approaches and principles of these while understanding the differences and particulars that the Internet of Things poses. The **framework for sustainable interoperability in Internet of Things applications** needs (at least) to address the following aspects:

- Management of Interoperability in the IoT: In order to correctly support interoperability in the Internet of Things one needs to efficiently and effectively manage interoperability resources. **What then needs to be managed, to what extent and how, in respect to interoperability in the Internet of Things?**

- Dynamic Interoperability Technologies for the IoT: In order for interoperability to endure in the complex IoT environment, one needs to permit Things to enter and dynamically interoperate without the need of being remanufactured. Then, what approaches and methods to create dynamic interoperability in IoT?

- Measurement of Interoperability in the IoT: In order to properly manage and execute interoperability in the IoT it is needs to quantify and/or qualify interoperability itself. As Lord Kelvin stated: "If one can not measure it, one can not improve it". Then, **what methods and techniques to provide an adequate measurement of Interoperability in the Internet of Things?**

- Interaction and integration of IoT in the global Internet: IPv6 integration, global interoperability, IoT-Cloud integration, etc. In other words, how to bridge billion of smart things globally, while respecting their specific constraints.

3.10.4.2 Technical IoT Interoperability

There are different areas on interoperability such as at least four areas on technical interoperability, syntactic, semantic interoperability and organizational interoperability. **Technical Interoperability** is usually associated with hardware/software components, systems and platforms that enable machine-to-machine communication to take place. This kind of interoperability is often centred on (communication) protocols and the infrastructure needed for those protocols to operate and we need to pay a specific attention as many protocols are developed within SDOs and therefore it will require market proof approach to validate and implement these protocols leading to have true interoperable and global IoT products.

Validation

Validation is an important aspect of interoperability (also in the Internet of Things). Testing and Validation provide the assurance that interoperability methods, protocols, etc. can cope with the specific nature and requirements of the Internet of Things.

The main way, among others, is to provide **efficient and accurate test suites and associated interoperability testing methodology** (with associated test description/coding languages) that help in testing thoroughly both the underlying protocols used by interconnected things / machines / smart objects and the embedded services / applications. The testing features and facilities need to become build into the design and deployment process, as the conditions of communication means, object/things availability and accessibility may change over time or location.

It is really important that these new testing methods consider the real context of future communicating systems where these objects will be deployed. Indeed, contrary to most of the existing testing methods, interconnected things / machines / smart objects in the IoT are naturally distributed. As they are distributed, the usual and classical approach of a single centralized testing system dealing with all these components and the test execution is no more applicable. The distributed nature of the tested components imposes to move towards distributed testing methods. To be more confident in the real interoperability of these components when they will be deployed in real networks, testing has to be done in a (close to) real operational environment. In this context of IoT where objects are connected through radio links, communicating environment may be unreliable and non-controllable if don't address seriously interoperability testing challenges with the same intensity and complexity of the IoT research itself. **Research in IoT challenges leads to IoT validation and interoperability challenges**.

3.11 IoT Protocols Convergence

In order to use the full potential of IoT paradigm the interconnected devices need to communicate using lightweight protocols that don't require extensive use of CPU resources. C, Java, MQTT, Python and some scripting languages are the preferable choices used by IoT applications. The IoT nodes use separate IoT gateways if there is needed protocol conversion, database storage, or decision making in order to supplement the low-intelligence node.

One of the most important aspects for a convergence protocol that support information exchange between domains, is the ability to convey the information (data) contained in a particular domain to other domains. This section provides an overview of the existing data exchange protocols that can be applied for data exchange among various domains.

Today there are two dominant architectures for data exchange protocols; bus-based, and broker-based. In the broker-based architecture, the broker

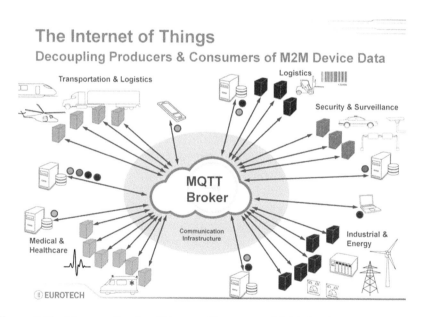

Figure 3.49 Message Queuing Telemetry Transport publish/subscribe protocol used to implement IoT and M2M applications (Source: Eurotech)

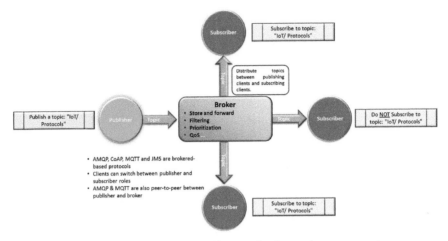

Figure 3.50   Broker based architecture for data exchange protocols

controls the distribution of the information. For example, it stores, forwards, filters and prioritizes publish requests from the publisher (the source of the information) client to the subscriber (the consumer of the information) clients. Clients switch between publisher and subscriber roles depending on their objectives. Examples of broker –based protocols include Advanced Message Queuing Protocol (AMPQ), Constrained Applications Protocol (CoAP), Message Queue Telemetry Transport (MQTT) and Java Message Service API (JMS).

In the bus-based architecture, clients publish messages for a specific topic which are directly delivered to the subscribers of that topic. There is no centralized broker or broker-based services. Examples of bus-based protocols include Data Distribution Service (DDS), Representational State Transfer (REST) and Extensible Messaging and Presence Protocol (XMPP).

Another important way to classify these protocols is whether they are message-centric or data-centric. Message centric protocols such as AMQP, MQTT, JMS and REST focus on the delivery of the message to the intended recipient(s), regardless of the data payload it contains. A data-centric protocol such as DDS, CoAP and XMPP focus on delivering the data and assumes the data is understood by the receiver. Middleware understands the data and ensures that the subscribers have a synchronized and consistent view of the data.

Yet another fundamental aspect of these protocols is whether it is web-based like CoAP or application-based such as with XMPP, and AMQP. These

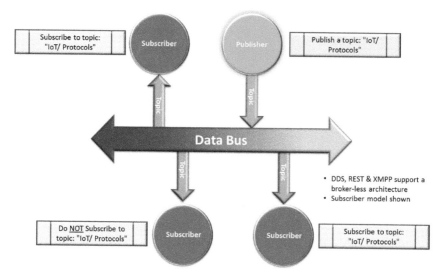

Figure 3.51 Bus-based architecture for data exchange protocols

aspects have fundamental effect on the environment, performance and tools available for implementers.

The following sections describe the example protocols in more detail, [31–33].

3.11.1 Message Queue Telemetry Transport (MQTT)

MQTT is an open-sourced protocol for passing messages between multiple clients through a central broker. It was designed to be simple and easy to implement. The MQTT architecture is broker-based, and uses long-lived outgoing TCP connection to the broker. MQTT also supports hierarchical topics (e.g., "subject/sub-subject/sub-sub-subject") file system structure.

MQTT can be used for two-way communications over unreliable networks where cost per transmitted bit is comparatively high. It is also compatible with low power consumption devices. The protocol is light-weight (simple) and therefore well suited for constrained environments. MQTT has a mechanism for asynchronous communication and for communicating disconnect messages when a device has disconnected. The most recent message can also be stored and forwarded. Multiple versions of MQTT are available to address specific limitations.

With MQTT, only partial interoperability between publishers and subscribers can be guaranteed because the meaning of data is not negotiated. Clients must know message format up-front. In addition, it does not support labeling messages with types or metadata. MQTT may include large topic strings that may not be suitable for small packet size of some transport protocols such as IEEE 802.15.4 without using MQTT-SN. MQTT may require EXI (Efficient XML Interchange) to compress the message length that could reduce communication efficiency.

TCP may negatively affect the network efficiency as the number of nodes (connection to the broker) increases. If the number of nodes is greater than a thousand, poor performance and complexity may also result because automatic/dynamic discovery is not supported in MQTT.

Because the protocol was designed to be simple, users must decide whether it is too simple and susceptible to potential hacking.

3.11.2 Constrained Applications Protocol (CoAP)

CoAP is an internet-based client/server model document transfer protocol similar to HTTP but designed for constrained devices. A sensor is typically a "server" of information and the "client" the consumer who can also alter states. It supports a one-to-one protocol for transferring state information between client and server.

CoAP utilizes User Datagram Protocol (UDP), and supports broadcast and multicast addressing. It does not support TCP. CoAP communication is through connectionless datagrams, and can be used on top of SMS and other packet-based communications protocols.

CoAP supports content negotiation and discovery, allowing devices to probe each other to find ways to exchange data. CoAP was designed for interoperability with the web (including HTTP and RESTful protocols), and supports asynchronous communications. The small packets are easy to generate. CoAP supports "observing" resource state changes as they occur so it is best suited to a state-transfer model, not purely an event-based model. CoAP supports a means for resource discovery.

UDP may be easier to implement in microcontrollers than TCP, but the security tools used for TCP (SSL/TLS) are not available in UDP. Datagram Transport Layer Security (DTLS) can be used instead. In addition, system issues such as the amount of support required for HTTP, Tunneling and Port Forwarding in NAT environments needs to be evaluated.

3.11.3 Advanced Message Queuing Protocol (AMQP)

AMQP is an application layer message-centric brokered protocol that emerged from the financial sector with the objective of replacing proprietary and non-interoperable messaging systems. The key features of AMQP are message orientation, queuing, routing (including point-to-point and publish-and-subscribe), reliability and security. Discovery is done via the broker.

It provides flow controlled, message-oriented communication with message-delivery guarantees such as at-most-once (where each message is delivered once or never), at-least-once (where each message is certain to be delivered, but may do so multiple times) and exactly-once (where the message will always certainly arrive and do so only once), and authentication and/or encryption based on SASL and/or TLS. It assumes an underlying reliable transport layer protocol such as Transmission Control Protocol (TCP) using SSL/TLS, [30].

AMQP mandates the behavior of the messaging provider and client to the extent that implementations from different vendors are truly interoperable. Previous attempts to standardize middleware have happened at the API level (e.g. JMS) and thus did not ensure interoperability. Unlike JMS, which merely defines an API, AMQP is a wire-protocol. Consequently any product that can create and interpret messages that conform to this data format can interoperate with any other compliant implementation irrespective of the programming language, [30].

Support for more than a thousand nodes may result in poor performance and increased complexity.

3.11.4 Java Message Service API (JMS)

JMS is a message oriented middleware API for creating, reading, sending, receiving messages between two or more clients, based on the Java Enterprise Edition. It was meant to separate application and transport layer functions and allows the communications between different components of a distributed application to be loosely coupled, reliable and asynchronous over TCP/IP.

JMS supports both the point to point and publish/subscribe models using message queuing, and durable subscriptions (i.e., store and forward topics to subscribers when they "log in"). Subscription control is through topics and queues with message filtering. Discovery is via the broker (server). The same Java classes can be used to communicate with different JMS providers by using the Java Naming and Directory interface for the desired provider.

When considering JMS API, keep in mind that it cannot guarantee interoperability between producers and consumers using different JMS implementations. Also, systems with more than a thousand nodes may result in poor performance and increased complexity.

3.11.5 Data Distribution Service (DDS)

DDS is a data-centric middleware language used to enable scalable, real-time, dependable high performance and interoperable data exchanges. The original target applications were financial trading, air traffic control, smart grid management and other big data, mission critical applications.

It is a decentralized broker-less protocol with direct peer-to-peer communications between publishers and subscribers and was designed to be language and operating system independent. DDS sends and receives data, events, and command information on top of UDP but can also run over other transports such as IP Multicast, TCP/IP, shared memory etc. DDS supports real-time many-to-many managed connectivity and also supports automatic discovery.

Applications using DDS for communications are decoupled and do not require intervention from the user applications, which can simplify complex network programming. QoS parameters that are used to configure its auto-discovery mechanisms are setup one time. DSS automatically handles hot-swapping redundant publishers if the primary publisher fails. Subscription control is via partitions and topics with message filtering.

DDS Security specification is still pending. Implementers should be aware that DSS needs DSSI ("wire-protocol") to make sure all implementations can interoperate.

DSS is available commercially and a version of it has been made "open" in as much as a "public" version is available.

3.11.6 Representational State Transfer (REST)

REST is a language and operating system independent architecture for designing network applications using simple HTTP to connect between machines. It was designed as a lightweight point-to-point, stateless client/server, cacheable protocol for simple client/server (request/reply) communications from devices to the cloud over TCP/IP.

Use of stateless model supported by HTTP and can simplify server design and can easily be used in the presence of firewalls, but may result in the need for additional information exchange. It does not support Cookies or asynchronous, loosely coupled publish-and-subscribe message exchanges.

Support for systems with more than a thousand nodes may result in poor performance and complexity.

3.11.7 Extensible Messaging and Presence Protocol (XMPP)

XMPP is a communications protocol for message oriented middleware based on XML (formally "Jabber"). It is a brokerless decentralized client-server (as previously defined) model and is used by text messaging applications. It is near real-time and massively scalable to hundreds of thousands of nodes. Binary data must be base64 encoded before it can be transmitted in-band.

It is useful for devices with large and potentially complicated traffic, and where extra security is required. For example, it can be used to isolate security to between applications rather than to rely on TCP or the web. The users or devices (servers) can keep control through preference settings.

New extensions being added to enhance its application to the IoT, including Service Discovery (XEP-0030), Concentrators for connecting legacy sensors and devices (XEP-0325), SensorData (XEP-0323), and Control (XEP-0322) and the Transport of XMPP over HTTM (XP-0124).

3.12 Discussion

The Internet of Things will grow to 26 billion units (without considering PCs, tablets and smartphones) installed in 2020 representing an almost 30-fold increase from 0.9 billion in 2009. IoT product and service suppliers will generate incremental revenue exceeding $300 billion, mostly in services, in 2020. It will result in $1.9 trillion in global economic value-add through sales into diverse end markets. Due to the low cost of adding IoT capability to consumer products, it is expected that "ghost" devices with unused connectivity will be common. This will be a combination of products that have the capability built in but require software to "activate" it and products with IoT functionality that customers do not actively leverage. In addition, enterprises will make extensive use of IoT technology, and there will be a wide range of products sold into various markets, such as advanced medical devices; factory automation sensors and applications in industrial robotics; sensor motes for increased agricultural yield; and automotive sensors and infrastructure integrity monitoring systems for diverse areas, such as road and railway transportation, water distribution and electrical transmission.

By 2020, component costs will have come down to the point that connectivity will become a standard feature, even for processors costing less than $1. This opens up the possibility of connecting just about anything, from the very simple to the very complex, to offer remote control, monitoring and sensing and it is expected that the variety of devices offered to explode [77].

The IoT encompasses sensor, actuators, electronic processing, microcontrollers, embedded software, communications services and information services associated with the things.

The economic value added at the European and global level is significant across sectors in 2020. The IoT applications are still implemented by the different industrial verticals with a high adoption in manufacturing, healthcare and home/buildings.

IoT will also facilitate new business models based on the real-time data acquired by billions of sensor nodes. This will push for development of advances sensor, nanoelectronics, computing, network and cloud technologies and will lead to value creation in utilities, energy, smart building technology, transportation and agriculture.

Acknowledgments

The IoT European Research Cluster - European Research Cluster on the Internet of Things (IERC) maintains its Strategic Research and Innovation Agenda (SRA), taking into account its experiences and the results from the on-going exchange among European and international experts.

The present document builds on the 2010, 2011, 2012, and 2013 Strategic Research and Innovation Agendas and presents the research fields and an updated roadmap on future R&D from 2015 to 2020 and beyond 2020.

The IoT European Research Cluster SRA is part of a continuous IoT community dialogue supported by the European Commission (EC) DG Connect – Communications Networks, Content and Technology, E1 - Network technologies Unit for the European and international IoT stakeholders. The result is a lively document that is updated every year with expert feedback from on-going and future projects financed by the EC. Many colleagues have assisted over the last few years with their views on the Internet of Things Strategic Research and Innovation agenda document. Their contributions are gratefully acknowledged.

Table 3.1 Future Technological Developments

Development	2015–2020	Beyond 2020
Identification Technology	• Identity management • Open framework for the IoT • Soft Identities • Semantics • Privacy awareness	"Thing/Object DNA" identifier
Internet of Things Architecture Technology	Network of networks architectures • IoT architecture developments • Adaptive, context based architectures • Self-* properties	Cognitive architectures • Experimental architectures
Internet of Things Infrastructure	Cross domain application deployment • Integrated IoT infrastructures • Multi application infrastructures • Multi provider infrastructures	Global, general purpose IoT infrastructures • Global discovery mechanism
Internet of Things Applications	Configurable IoT devices • IoT in food/water production and tracing • IoT in manufacturing industry • IoT in industrial lifelong service and maintenance • IoT device with strong processing and analytics capabilities • Application capable of handling heterogeneous high capability data collection an d processing infrastructures	IoT information open market
Communication Technology	Wide spectrum and spectrum aware protocols • Ultra low power chip sets • On chip antennas • Millimeter wave single chips • Ultra low power single chip radios • Ultra low power system on chip	Unified protocol over wide spectrum • Multi-functional reconfigurable chips
Network Technology	Network context awareness • Self aware and self organizing networks • Sensor network location transparency • IPv6- enabled scalability	Network cognition • Self-learning, self-repairing networks • Ubiquitous IPv6-based IoT deployment

(Continued)

Table 3.1   Continued

Development	2015–2020	Beyond 2020
Software and algorithms	Goal oriented software • Distributed intelligence, problem solving • Things-to-Things collaboration environments • IoT complex data analysis • IoT intelligent data visualization • Hybrid IoT and industrial automation systems	User oriented software • The invisible IoT • Easy-to-deploy IoT sw • Things-to-Humans collaboration • IoT 4 All • User-centric IoT
Hardware	Smart sensors (bio-chemical) • More sensors and actuators (tiny sensors) • Sensor integration with NFC • Home printable RFID tags	Nano-technology and new materials
Data and Signal Processing Technology	Context aware data processing and data responses • Energy, frequency spectrum aware data processing	Cognitive processing and optimisation
Discovery and Search Engine Technologies	Automatic route tagging and identification management centres • Semantic discovery of sensors and sensor data	Cognitive search engines • Autonomous search engines
Power and Energy Storage Technologies	Energy harvesting (biological, chemical, induction) • Power generation in harsh environments • Energy recycling • Long range wireless power • Wireless power	Biodegradable batteries • Nano-power processing unit
Security, Privacy & Trust Technologies	User centric context-aware privacy and privacy policies • Privacy aware data processing • Security and privacy profiles selection based on security and privacy needs • Privacy needs automatic evaluation • Context centric security • Homomorphic Encryption • Searchable Encryption • Protection mechanisms for IoT DoS/DdoS attacks	Self adaptive security mechanisms and protocols • Self-managed secure IoT

Table 3.1   (Continued)

Development	2015–2020	Beyond 2020
Material Technology	SiC, GaN • Improved/new semiconductor manufacturing processes/technologies for higher temperature ranges	Diamond • Graphen
Interoperability	Optimized and market proof interoperability approaches used • Interoperability under stress as market grows • Cost of interoperability reduced • Several successful certification programmes in place	Automated self-adaptable and agile interoperability
Standardisation	IoT standardization refinement • M2M standardization as part of IoT standardisation • Standards for cross interoperability with heterogeneous networks • IoT data and information sharing	Standards for autonomic communication protocols

Table 3.2   Internet of Things Research Needs

Research needs	2015–2020	Beyond 2020
Identification Technology	Convergence of IP and IDs and addressing scheme • Unique ID • Multiple IDs for specific cases • Extend the ID concept (more than ID number) • Electro Magnetic Identification – EMID	Multi methods – one ID
IoT Architecture	Internet (Internet of Things) (global scale applications, global interoperability, many trillions of things)	
Internet of Things Infrastructure	Application domain-independent abstractions & functionality • Cross-domain integration and management • Large-scale deployment of infrastructure • Context-aware adaptation of operation	Self management and configuration

(*Continued*)

Table 3.2   (Continued)

Research needs	2015–2020	Beyond 2020
Internet of Things Applications	IoT information open market • Standardization of APIs • IoT device with strong processing and analytics capabilities • Ad-hoc deployable and configurable networks for industrial use • Mobile IoT applications for IoT industrial operation and service/maintenance • Mobile IoT applications for IoT industrial operation and service/maintenance • Fully integrated and interacting IoT applications for industrial use	Building and deployment of public IoT infrastructure with open APIs and underlying business models • Mobile applications with bio-IoT-human interaction
SOA Software Services for IoT	Quality of Information and IoT service reliability • Highly distributed IoT processes • Semi-automatic process analysis and distribution	Fully autonomous IoT devices
Internet of Things Architecture Technology	Code in tags to be executed in the tag or in trusted readers • Global applications • Adaptive coverage • Universal authentication of objects • Graceful recovery of tags following power loss • More memory • Less energy consumption • 3-D real time location/position embedded systems	Intelligent and collaborative functions • Object intelligence • Context awareness • Cooperative position cyber-physical systems
Communication Technology	Longer range (higher frequencies – tenths of GHz) • Protocols for interoperability • On chip networks and multi standard RF architectures • Multi-protocol chips • Gateway convergence	Self configuring, protocol seamless networks

Network Technology	• Hybrid network technologies convergence • 5G developments • Collision-resistant algorithms • Plug and play tags • Self repairing tags Grid/Cloud network • Software defined networks • Service based network • Multi authentication • Integrated/universal authentication • Brokering of data through market mechanisms • Scalability enablers • IPv6-based networks for smart cities	Need based network • Internet of **Everything** • Robust security based on a combination of ID metrics • Autonomous systems for non stop information technology service • Global European IPv6-based Internet of Everything
Software and algorithms	Self management and control • Micro operating systems • Context aware business event generation • Interoperable ontologies of business events • Scalable autonomous software Evolving software • Self reusable software • Autonomous things: • Self configurable • Self healing • Self management • Platform for object intelligence	Self generating "molecular" software • Context aware software
Hardware Devices	Polymer based memory • Ultra low power **EPROM/FRAM** • Molecular sensors • Autonomous circuits • Transparent displays • Interacting tags • Collaborative tags • Heterogeneous integration • Self powering sensors • Low cost modular devices	Biodegradable antennas • Autonomous "bee" type devices

(Continued)

Table 3.2 (Continued)

Research needs	2015–2020	Beyond 2020
	• Ultra low power circuits • Electronic paper • Nano power processing units • Silent Tags • Biodegradable antennae	
Hardware Systems, Circuits and Architectures	Multi protocol front ends • Ultra low cost chips with security • Collision free air to air protocol • Minimum energy protocols • Multi-band, multi-mode wireless sensor architectures implementations • Adaptive architectures • Reconfigurable wireless systems • Changing and adapting functionalities to the environments • Micro readers with multi standard protocols for reading sensor and actuator data • Distributed memory and processing Low cost modular devices • Protocols correct by construction	Heterogeneous architectures • "Fluid" systems, continuously changing and adapting
Data and Signal Processing Technology	Common sensor ontologies (cross domain) • Distributed energy efficient data processing • Autonomous computing • Tera scale computing	Cognitive computing
Discovery and Search Engine Technologies	Scalable Discovery services for connecting things with services while respecting security, privacy and confidentiality • "Search Engine" for Things • IoT Browser • Multiple identities per object • On demand service discovery/integration • Universal authentication	Cognitive registries

Power and Energy Storage Technologies	Paper based batteries • Wireless power everywhere, anytime • Photovoltaic cells everywhere • Energy harvesting • Power generation for harsh environments	Biodegradable batteries
Interoperability	Dynamic and adaptable interoperability for technical and semantic areas • Open platform for IoT validation	Self-adaptable and agile interoperability approaches
Security, Privacy & Trust Technologies	Low cost, secure and high performance identification/authentication devices • Access control and accounting schemes for IoT • General attack detection and recovery/resilience for IoT • Cyber Security Situation Awareness for IoT • Context based security activation algorithms • Service triggered security • Context-aware devices • Object intelligence Decentralised self configuring methods for trust establishment • Novel methods to assess trust in people, devices and data • Location privacy preservation • Personal information protection from inference and observation • Trust Negotiation	Cognitive security systems • Self-managed secure IoT • Decentralised approaches to privacy by information localisation
Governance (legal aspects)	Legal framework for transparency of IoT bodies and organizations	Adoption of clear European norms/standards regarding Privacy and Security for IoT

(Continued)

Table 3.2   (Continued)

Research needs	2015–2020	Beyond 2020
Economic	• Privacy knowledge base and development privacy standards Business cases and value chains for IoT • Emergence of IoT in different industrial sectors	
Material Technology	Carbon nanotube • Conducting Polymers and semiconducting polymers and molecules • Modular manufacturing techniques	Graphen

List of Contributors

Abdur Rahim Biswas, IT, create-net, iCore
Alessandro Bassi, FR, Bassi Consulting, IoT-A
Ali Rezafard, IE, Afilias, EPCglobal Data Discovery JRG
Amine Houyou, DE, SIEMENS, IoT@Work
Antonio Skarmeta, SP, University of Murcia, IoT6
Carlos Agostinho, PT, UNINOVA
Carlo Maria Medaglia, IT, University of Rome 'Sapienza', IoT-A
César Viho, FR, Probe-IT
Claudio Pastrone, IT, ISMB, ebbits, ALMANAC
Daniel Thiemert, UK, University of Reading, HYDRA
David Simplot-Ryl, FR, INRIA/ERCIM, ASPIRE
Elias Tragos, GR, FORTH, RERUM
Eric Mercier, FR, CEA-Leti
Erik Berg, NO, Telenor, IoT-I
Francesco Sottile, IT, ISMB, BUTLER
Franck Le Gall, FR, Inno, PROBE-IT, BUTLER
François Carrez, GB, IoT-I
Frederic Thiesse, CH, University of St. Gallen, Auto-ID Lab
Friedbert Berens, LU, FB Consulting S.à r.l, BUTLER
Gary Steri, IT, EC, JRC
Gianmarco Baldini, IT, EC, JRC
Giuseppe Abreu, DE, Jacobs University Bremen, BUTLER
Ghislain Despesse, FR, CEA-Leti

Hanne Grindvoll, NO, SINTEF ICT
Harald Sundmaeker, DE, ATB GmbH, SmartAgriFood, CuteLoop
Henri Barthel, BE, GS1 Global
Igor Nai Fovino, IT, EC, JRC
Jan Höller, SE, EAB
Jens-Matthias Bohli, DE, NEC
John Soldatos, GR, Athens Information Technology, ASPIRE, OpenIoT
Jose-Antonio, Jimenez Holgado, ES, TID
Klaus Moessner, UK, UNIS, IoT.est
Kostas Kalaboukas, GR, SingularLogic, EURIDICE
Latif Ladid, LU, UL, IPv6 Forum
Levent Gürgen, FR, CEA-Leti
Luis Muñoz, ES, Universidad De Cantabria
Manfred Hauswirth, IE, DERI, OpenIoT, VITAL
Marco Carugi, IT, ITU-T, ZTE
Marilyn Arndt, FR, Orange
Mario Hoffmann, DE, Fraunhofer-Institute SIT, HYDRA
Markus Eisenhauer, DE, Fraunhofer-FIT, HYDRA, ebbits
Markus Gruber, DE, ALUD
Martin Bauer, DE, NEC, IoT-A
Martin Serrano, IE, DERI, OpenIoT
Maurizio Spirito, IT, Istituto Superiore Mario Boella, , ebbits, ALMANAC
Maarten Botterman, NL, GNKS, SMART-ACTION
Nicolaie L. Fantana, DE, ABB AG
Nikos Kefalakis, GR, Athens Information Technology, OpenIoT
Paolo Medagliani, FR, Thales Communications & Security, CALYPSO
Payam Barnaghi, UK, UNIS, IoT.est
Philippe Cousin, FR, easy global market, PROBE-IT,
Raffaele Giaffreda, IT, CNET, iCore
Ricardo Neisse, IT, EC, JRC
Richard Egan, UK, TRT
Rolf Weber, CH, UZH
Sébastien Boissseau, FR, CEA-Leti
Sébastien Ziegler, CH, Mandat International, IoT6
Sergio Gusmeroli, IT, TXT e-solutions,
Stefan Fisher, DE, UZL
Stefano Severi, DE, Jacobs University Bremen, BUTLER
Srdjan Krco, RS, DunavNET,, IoT-I, SOCIOTAL
Sönke Nommensen, DE, UZL, SmartSantander

Trevor Peirce, BE, CASAGRAS2
Veronica Gutierrez Polidura, ES, Universidad De Cantabria
Vincent Berg, FR, CEA-Leti
Vlasios Tsiatsis, SE, EAB
Wolfgang König, DE, ALUD
Wolfgang Templ, DE, ALUD

Contributing Projects and Initiatives

ASPIRE, BRIDGE, CASCADAS, CONFIDENCE, CuteLoop, DACAR, ebbits, ARTEMIS, ENIAC, EPoSS, EU-IFM, EURIDICE, GRIFS, HYDRA, IMS2020, Indisputable Key, iSURF, LEAPFROG, PEARS Feasibility, PrimeLife, RACE networkRFID, SMART, StoLPaN, SToP, TraSer, WALTER, IoT-A, IoT@Work, ELLIOT, SPRINT, NEFFICS, IoT-I, CASAGRAS2, eDiana, OpenIoT, IoT6, iCore PROBE-IT, BUTLER, IoT-est, SmartAgri-Food, ALMANAC, CITYPULSE,COSMOS,CLOUT, RERUM, SMARTIE, SOCIOTAL, VITAL

List of Abbreviations and Acronyms

Acronym	Meaning
3GPP3GPP	3rd Generation Partnership Project
AAL	Ambient Assisted Living
ACID	Atomicity, Consistency, Isolation, Durability
ACL	Access Control List
AMR	Automatic Meter Reading Technology
API	Application Programming Interface
AWARENESS	EU FP7 coordination action Self-Awareness in Autonomic Systems
BACnet	Communications protocol for building automation and control networks
BAN	Body Area Network
BDI	Belief-Desire-Intention architecture or approach
Bluetooth	Proprietary short range open wireless technology standard
BPM	Business process modelling
BPMN	Business Process Model and Notation
BUTLER	EU FP7 research project uBiquitous, secUre inTernet of things with Location and contExt-awaReness
CAGR	Compound annual growth rate
CE	Council of Europe
CENCEN	Comité Européen de Normalisation

CENELEC	Comité Européen de Normalisation Électrotechnique
CEO	Chief executive officer
CEP	Complex Event Processing
CMOS	Complementary metal-oxide-semiconductor
CSS	Chirp Spread Spectrum
D1.3	Deliverable 1.3
DATEX-II	Standard for data exchange involving traffic centres
DCA	Data Collection and Analysis
DNS	Domain Name System
DoS/DDOS	Denial of service attack Distributed denial of service attack
EC	European Commission
eCall	eCall – eSafety Support A European Commission funded project, coordinated by ERTICO-ITS Europe
EDA	Event Driven Architecture
EH	Energy harvesting
EMF	Electromagnetic Field
ERTICO-ITS	Multi-sector, public / private partnership for intelligent transport systems and services for Europe
ESOs	European Standards Organisations
ESP	Event Stream Processing
ETSI	European Telecommunications Standards Institute
EU	European Union
Exabytes	10^{18} bytes
FI	Future Internet
FI PPP	Future Internet Public Private Partnership programme
FIA	Future Internet Assembly
FIS 2008	Future Internet Symposium 2008
F-ONS	Federated Object Naming Service
FP7	Framework Programme 7
FTP	File Transfer Protocol
GFC	Global Certification Forum
GreenTouch	Consortium of ICT research experts
GS1	Global Standards Organization
Hadoop	Project developing open-source software for reliable, scalable, distributed computing
IAB	Internet Architecture Board
IBM	International Business Machines Corporation
ICAC	International Conference on Autonomic Computing
ICANN	Internet Corporation for Assigned Name and Numbers
ICT	Information and Communication Technologies
iCore	EU research project Empowering IoT through cognitive technologies
IERC	European Research Cluster for the Internet of Things
IETF	Internet Engineering Task Force
INSPIRE	Infrastructure for Spatial Information in the European Community

IoE	Internet of EnergyInternet of Energy
IoM	Internet of MediaInternet of Media
IoP	Internet of PersonsInternet of Persons, Internet of PeopleInternet of People
IoS	Internet of ServicesInternet of Services
IoT	Internet of Things
IoT6	EU FP7 research project Universal integration of the Internet of Things through an IPv6-based service oriented architecture enabling heterogeneous components interoperability
IoT-A	Internet of Things ArchitectureInternet of Things Architecture
IoT-A	Internet of Things ArchitectureInternet of Things Architecture
IoT-est	EU ICT FP7 research project Internet of Things environment for service creation and testing
IoT-i	Internet of Things Initiative
IoV	Internet of Vehicles
IP	Internet Protocol
IPSO Alliance	Organization promoting the Internet Protocol (IP) for Smart Object communications
IPv6	Internet Protocol version 6
ISO 19136	Geographic information, Geography Mark-up Language, ISO Standard
IST	Intelligent Transportation System
KNX	Standardized, OSI-based network communications protocol for intelligent buildings
LNCS	Lecture Notes in Computer Science
LOD	Linked Open Data Cloud
LTE	Long Term Evolution
M2M	Machine to Machine
MAC	Media Access Control data communication protocol sub-layer
MAPE-K	Model for autonomic systems: Monitor, Analyse, Plan, Execute in interaction with a Knowledge base
makeSense	EU FP7 research project on Easy Programming of Integrated Wireless Sensors
MB	Megabyte
MIT	Massachusetts Institute of Technology
MPP	Massively parallel processing
NIEHS	National Institute of Environmental Health Sciences
NFC	Near Field Communication
NoSQL	not only SQL – a broad class of database management systems

OASIS	Organisation for the Advancement of Structured Information Standards
OEM	Original equipment manufacturer
OGC	Open Geospatial Consortium
OMG	Object Management Group
OpenIoT	EU FP7 research project Part of the Future Internet public private partnership Open source blueprint for large scale self-organizing cloud environments for IoT applications
Outsmart	EU project Provisioning of urban/regional smart services and business models enabled by the Future Internet
PAN	Personal Area Network
PET	Privacy Enhancing Technologies
Petabytes	10^{15} byte
PHY	Physical layer of the OSI model
PIPES	Public infrastructure for processing and exploring streams
PKI	Public key infrastructure
PPP	Public-private partnership
Probe-IT	EU ICT-FP7 research project Pursuing roadmaps and benchmarks for the Internet of Things
PSI	Public Sector Information
PV	Photo Voltaic
QoI	Quality of Information
RF	Radio frequency
RFID	Radio-frequency identification
SASO	IEEE international conferences on Self-Adaptive and Self-Organizing Systems
SDO	Standard Developing Organization
SEAMS	International Symposium on Software Engineering for Adaptive and Self-Managing Systems
SENSEI	EU FP7 research project Integrating the physical with the digital world of the network of the future
SIG	Special Interest Group
SLA	Service-level agreement / Software license agreement
SmartAgriFood	EU ICT FP7 research project Smart Food and Agribusiness: Future Internet for safe and healthy food from farm to fork
SmartSantander	EU ICT FP7 research project Future Internet research and experimentation
SOA	Service Oriented Approach
SON	Self Organising Networks
SSW	Semantic Sensor Web
SRA	Strategic Research Agenda
SRIA	Strategic Research and Innovation Agenda
SRA2010	Strategic Research Agenda 2010
SWE	Sensor Web Enablement
TC	Technical Committee

TTCN-3	Testing and Test Control Notation version 3
USDL	Unified Service Description Language
UWB	Ultra-wideband
W3C	World Wide Web Consortium
WS&AN	Wireless sensor and actuator networks
WSN	Wireless sensor network
WS-BPEL	Web Services Business Process Execution Language
Zettabytes	10^{21} byte
ZigBee	Low-cost, low-power wireless mesh network standard based on IEEE 802.15.4

References

[1] NFC Forum, online at http://nfc-forum.org

[2] METIS, Mobile and wireless communications Enablers for the Twenty-twenty (2020) Information Society, online at https://www.metis2020.com/

[3] Wemme, L., "NFC: Global Promise and Progress", NFC Forum, 22.01.2014, online at http://nfc-forum.org/wp-content/uploads/2014/01/Omnicard_Wemme_2014_website.pdf

[4] Bluetooth Special Interest Group, online at https://www.bluetooth.org/en-us/members/about-sig

[5] Bluetooth Developer Portal, online at https://developer.bluetooth.org/Pages/default.aspx

[6] Bluetooth, online at http://www.bluetooth.com

[7] ANT+, online at http://www.thisisant.com/

[8] ANT, "Message Protocol and Usage rev.5.0", online at http://www.thisisant.com/developer/resources/downloads#documents_tab

[9] ANT, "FIT2 Fitness Module Datasheet", online at http://www.thisisant.com/developer/resources/downloads#documents_tab

[10] Wi-Fi Alliance, online at http://www.wi-fi.org/

[11] Z-Wave alliance, online at http://www.z-wavealliance.org

[12] Pätz, C., "Smart lighting. How to develop Z-Wave Devices", EE Times europe LEDLighting, 04.10.2012, online at http://www.ledlighting-eetimes.com/en/how-to-develop-z-wave-devices.html?cmp_id=71&news_id=222908151

[13] KNX, online at http://www.knx.org/knx-en/knx/association

[14] European Editors, "Using Ultra-Low-Power Sub-GHz Wireless for Self-Powered Smart-Home Networks", 12.05.2013, online at

http://www.digikey.com/en-US/articles/techzone/2013/dec/using-ultra-low-power-sub-ghz-wireless-for-self-powered-smart-home-networks

[15] HART Communication Foundation, online at http://www.hartcomm .org

[16] Mouser Electronics, "Wireless Mesh Networking – Featured Wireless Mesh Networking Protocols", online at http://no.mouser.com/ applications/wireless_mesh_networking_protocols/

[17] IETF, online at https://www.ietf.org

[18] Bormann, C., "6LoWPAN Roadmap and Implementation Guide", 6LoWPAN Working Group, April 2013, http://tools.ietf.org/html/draft-bormann-6lowpan-roadmap-04

[19] Shelby, Z. and Bormann, C., "6LoWPAN: The Wireless Embedded Internet", Wiley, Great Britain, ISBN 9780470747995, 2009, online at http://elektro.upi.edu/pustaka.elektro/Wireless%20Sensor%20Network/ 6LoWPAN.pdf

[20] WiMAX Forum, online at http://www.wimaxforum.org

[21] A. Passemard, "The Internet of Things Protocol stack – from sensors to business value", online at http://entrepreneurshiptalk.wordpress.com/ 2014/01/29/the-internet-of-thing-protocol-stack-from-sensors-to-business-value/

[22] EnOcean Alliance, online at http://www.enocean-alliance.org/en/profile/

[23] EnOcean Wireless Standard, online at http://www.enocean.com

[24] EnOcean Alliance, "EnOcean Equipment Profiles (EEP)", Ver. 2.6, December 2013, online at http://www.enocean.com/en/home/

[25] DASH7 Alliance, online at http://www.dash7.org

[26] Maarten Weyn, "Dash7 Alliance Protocol Technical Presentation", December 2013, online at http://www.slideshare.net/MaartenWeyn1/ dash7-alliance-protocol-technical-presentation

[27] Visible Assets, Inc., "Rubee Technology", online at http://www.rubee .com/Techno/index.html

[28] Stevens, J., Weich, C., GilChrist, R., "RuBee (IEEE 1902.1) – The Physics Behind, Real-Time, High Security Wireless Asset Visibility Networks in Harsh Environments", online at http://www.rubee.com/White-SEC/RuBee-Security-080610.pdf

[29] RuBee Hardware, online at http://www.rubee.com/page2/Hard/index.html

[30] Foster, A., "A Comparison Between DDS, AMQP, MQTT, JMS, REST and CoAP", Version 1.4, January 2014, online at http://www.prismtech .com/sites/default/files/documents/MessagingComparsionJan2014 USROW_vfinal.pdf

[31] Elkstein, M., "Learn REST: A tutorial", online at http://rest.elkstein.org
[32] Jaffey, T., "MQTT and CoAP IoT Protocols.pdf", September 2013, online at https://docs.google.com/document/d/1_kTNkl84o_yoC56dzFfkYHo HuepINP3nDNokycXINXI/edit?usp=sharing&pli=1
[33] Puzanov, O., "IoT Protocol Wars: MQTT vs COAP vs XMPP", online at http://www.iotprimer.com/2013/11/iot-protocol-wars-mqtt-vscoap-vs-xmpp.html
[34] Home Gateway Initiative (HGI), online at www.homegatewayinitiative .org
[35] Artemis IoE project, online at www.artemis-ioe.eu
[36] Larios V.M., Robledo J.G., Gómez L., and Rincon R., "IEEE-GDL CCD Smart Buildings Introduction", online at http://smartcities.ieee.org/ images/files/ images/pdf/whitepaper_phi_smartbuildingsv6.pdf
[37] Analysys Mason, "Imagine an M2M world with 2.1 billion connected things", online at http://www.analysysmason.com/about-us/news/insight/M2M_forecast_Jan2011/
[38] Casaleggio Associati, "The Evolution of Internet of Things", February 2011, online at http://www.casaleggio.it/pubblicazioni/Focus_internet_of _things_v1.81%20-%20eng.pdf
[39] J. B., Kennedy, "When woman is boss, An interview with Nikola Tesla", in Colliers, January 30, 1926.
[40] M. Weiser, "The Computer for the 21st Century," *Scientific Am.*, Sept., 1991, pp. 94–104; reprinted in *IEEE Pervasive Computing*, Jan.-Mar. 2002, pp. 19–25."
[41] K. Ashton, "That 'Internet of Things' Thing", online at http://www. rfidjournal.com/article/view/4986, June 2009
[42] N. Gershenfeld, "When Things Start to Think", Holt Paperbacks, New York, 2000
[43] Raymond James & Associates, " The Internet of Things - A Study in Hype, Reality, Disruption, and Growth", online at http://sitic.org/wp-content/uploads/The-Internet-of-Things-A-Study-in-Hype-Reality-Disruption-and-Growth.pdf, January 2014.
[44] IDC, "Worldwide Internet of Things 2013–2020 Forecast: Billions of Things, Trillions of Dollars," Doc #: 243661, October 2013.
[45] N. Gershenfeld, R. Krikorian and D. Cohen, *Scientific Am.*, Sept., 2004
[46] World Economic Forum, "The Global Information Technology Report 2012 - Living in a Hyperconnected World" online at http://www3.weforum.org/docs/Global_IT_Report_2012.pdf
[47] "Key Enabling Technologies", Final Report of the HLG-KET, June 2011

[48] G. Bovet, A. Ridi and J. Hennebert, "Toward Web Enhanced Building Automation System", in Eds. N. Bessis and C. Dobre - Big Data and Internet of Things: A Roadmap for Smart Environments, ISBN: 978–3-319–05028-7, Studies in Computational Intelligence, Volume 546, 2014 pp. 259–283, online at http://hal.archives-ouvertes.fr/docs/00/97/35/10/PDF/BuildingsWoT.pdf

[49] International Technology Roadmap for Semiconductors, ITRS 2012 Update, online at http://www.itrs.net/Links/2012ITRS/2012Chapters /2012Overview.pdf

[50] W. Arden, M. Brillouët, P. Cogez, M. Graef, et al., "More than Moore" White Paper, online at http://www.itrs.net/Links/2010ITRS/IRC-ITRS-MtM-v2%203.pdf

[51] Frost & Sullivan "Mega Trends: Smart is the New Green" online at http://www.frost.com/prod/servlet/our-services-page.pag?mode=open &sid=230169625

[52] E. Savitz, "Gartner: 10 Critical Tech Trends For The Next Five Years" online at http://www.forbes.com/sites/ericsavitz/2012/10/22/gartner-10-critical-tech-trends-for-the-next-five-years/

[53] E. Savitz, "Gartner: Top 10 Strategic Technology Trends For 2013" online at http://www.forbes.com/sites/ericsavitz/2012/10/23/gartner-top-10-strategic-technology-trends-for-2013/

[54] P. High "Gartner: Top 10 Strategic Technology Trends For 2014" online at http://www.forbes.com/sites/peterhigh/2013/10/14/gartner-top-10-strategic-technology-trends-for-2014/#

[55] Platform INDUSTRIE 4.0 - Recommendations for implementing the strategic initiative industrie 4.0, Final report of the Industrie 4.0 Working Group, online at, ,http://www.acatech.de/fileadmin/user_upload /Baumstruktur_nach_Website/Acatech/root/de/Material_fuer_ Sonderseiten/Industrie_4.0/Final_report__Industrie_4.0_accessible.pdf, 2013

[56] Industrial Internet of Things (IoT) Advisory Service, ARC Advisory Group, online at, http://www.arcweb.com/services/pages/industrial-internet-of-things-service.aspx

[57] rtSOA - A Data Driven, Real Time Service Oriented Architecture for Industrial Manufacturing, online at http://www-db.in.tum.de/research/ projects/rtSOA/

[58] P. C. Evans and M. Annunziata, Industrial Internet: Pushing the Boundaries of Minds and Machines, General Electric Co., online

at http://files.gereports.com/wp-content/uploads/2012/11/ge-industrial-internet-vision-paper.pdf

[59] Cisco, "Securely Integrating the Cyber and Physical Worlds", online at http://www.cisco.com/web/solutions/trends/tech-radar/securing-the-iot.html

[60] NXT Cities, online at http://www.communicasia.com/wp-content/themes/cmma2014/images/img-nxtcities-large.jpg

[61] NXT Enterprises, online at http://www.communicasia.com/wp-content/themes/cmma2014/images/img-nxtenterprise-large.jpg

[62] NXT Connect, online at http://www.communicasia.com/wp-content/themes/cmma2014/images/img-nxtconnect-large.jpg

[63] H. Bauer, F. Grawert, and S. Schink, Semiconductors for wireless communications: Growth engine of the industry, online at www.mckinsey.com/

[64] L. Fretwell and P. Schottmiller, Cisco Presentation, online at http://www.cisco.com/assets/events/i/nrf-Internet_of_Everything_Whats_the_Art_of_the_Possible_in_Retail.pdf, January 2014.

[65] ITU-T, Internet of Things Global Standards Initiative, http://www.itu.int/en/ITU-T/gsi/iot/Pages/default.aspx

[66] International Telecommunication Union - ITU-T Y.2060 - (06/2012) – Next Generation Networks – Frameworks and functional architecture models - Overview of the Internet of things

[67] IEEE-SA - Enabling Consumer Connectivity Through Consensus Building, online at http://standardsinsight.com/ieee_company_detail/consensus-building

[68] O. Vermesan, P. Friess, P. Guillemin, S. Gusmeroli, et al., "Internet of Things Strategic Research Agenda", Chapter 2 in Internet of Things - Global Technological and Societal Trends, River Publishers, 2011, ISBN 978–87-92329–67-7

[69] O. Vermesan, P. Friess, P. Guillemin, H. Sundmaeker, et al., "Internet of Things Strategic Research and Innovation Agenda", Chapter 2 in Internet of Things – Converging Technologies for Smart Environments and Integrated Ecosystems, River Publishers, 2013, ISBN 978–87-92982–73-5

[70] SmartSantander, EU FP7 project, Future Internet Research and Experimentation, online at http://www.smartsantander.eu/

[71] Internet of Things Concept, online at http://xarxamobal.diba.cat/XGMSV/imatges/actualitat/iot.jpg

[72] H. Grindvoll, O. Vermesan, T. Crosbie, R. Bahr, et al., "A wireless sensor network for intelligent building energy management based on multi communication standards – a case study", ITcon Vol. 17, pg. 43–62, http://www.itcon.org/2012/3

[73] EU Research & Innovation, "Horizon 2020", The Framework Programme for Research and Innovation, online at http://ec.europa.eu/research/horizon2020/index_en.cfm

[74] Digital Agenda for Europe, European Commission, Digital Agenda 2010–2020 for Europe, online at http://ec.europa.eu/information_society/digital-agenda/index_en.htm

[75] S. Wilson, Deloitte Research, "Rising tide Exploring pathways to growth in the mobile semiconductor industry", 2013, online at http://dupress.com/articles/rising-tide-exploring-pathways-to-growth-in-the-mobile-semiconductor-industry/

[76] Gartner, "Hype Cycle for Emerging Technologies", 2011, online at http://www.gartner.com/it/page.jsp?id=1763814

[77] Gartner, 2013, online at http://www.gartner.com/newsroom/id/2636073

[78] K. Karimi and G. Atkinson, " What the Internet of Things (IoT) Needs to Become a Reality", White Paper, 2013, online at http://www.freescale.com/files/32bit/doc/white_paper/INTOTHNGSWP.pdf

[79] D. Evans, "The Internet of Things - How the Next Evolution of the Internet Is Changing Everything", CISCO White Paper, April 2011, online at http://www.cisco.com/web/about/ac79/docs/innov/IoT_IBSG_0411FINAL.pdf

[80] Freescale vision chip makes self-driving cars a bit more ordinary, online at http://www.cnet.com/news/freescale-vision-chip-makes-self-driving-cars-a-bit-more-ordinary/

[81] R. E. Hall, "The Vision of A Smart City" presented at the 2nd International Life Extension Technology Workshop Paris, France September 28, 2000 , online at http://www.crisismanagement.com.cn/templates/blue/down_list/llzt_zhcs/The%20Vision%20of%20A%20Smart%20City.pdf

[82] EU 2012. The ARTEMIS Embedded Computing Systems Initiative, October 2012 online at http://www.artemis-ju.eu/

[83] Foundations for Innovation in Cyber-Physical Systems, Workshop Report, NIST, 2013, online at http://www.nist.gov/el/upload/CPS-WorkshopReport-1–30-13-Final.pdf

[84] IERC – European Research Cluster on the Internet of Things, "Internet of Things - Pan European Research and Innovation Vision", October, 2011,

online at, http://www.theinternetofthings.eu/sites/default/files/Rob%20 van%20Kranenburg/IERC_IoT-Pan%20European%20Research%20and %20Innovation%20Vision_2011.pdf

[85] O. Vermesan, P. Friess, G. Woysch, P. Guillemin, S. Gusmeroli, et al., " Europe's IoT Stategic Research Agenda 2012", Chapter 2 in The Internet of Things 2012 New Horizons, Halifax, UK, 2012, ISBN 978 - 0 - 9553707 - 9 – 3

[86] SENSEI, EU FP7 project, *D1.4:* Business models and Value Creation, 2010, online at: http://www.ict-sensei.org.

[87] IoT-I, Internet of Things Initiative, FP7 EU project, 0nline at http://www.iot-i.eu

[88] Libelium, "50 Sensor Applications for a Smarter World", online at http://www.libelium.com/top_50_iot_sensor_applications_ranking#

[89] OUTSMART, FP7 EU project, part of the Future Internet Private Public Partnership, "OUTSMART - Provisioning of urban/regional smart services and business models enabled by the Future Internet", online at http://www.fi-ppp-outsmart.eu/en-uk/Pages/default.aspx

[90] BUTLER, FP7 EU project, online at http://www.iot-butler.eu/

[91] NXP Semiconductors N.V., "What's Next for Internet-Enabled Smart Lighting?", online at http://www.nxp.com/news/press-releases/2012/05/ whats-next-for-internet-enabled-smart-lighting.html

[92] J. Formo, M. Gårdman, and J. Laaksolahti, "Internet of things marries social media", in *Proceedings of the 13th International Conference on MobileHCI,* ACM, New York, NY, USA, pp. 753–755, 2011

[93] J. G. Breslin, S. Decker, and M. Hauswirth, et. al., "Integrating Social Networks and Sensor Networks", *W3C Workshop on the Future of Social Networking,* Barcelona, 15–16 January 2009

[94] M. Kirkpatrick, "The Era of Location-as-Platform Has Arrived", *ReadWriteWeb,* January 25, 2010

[95] F. Calabrese, K. Kloeckl, and C. Ratti (MIT), "WikiCity: Real-Time Location-Sensitive tools for the city", in *IEEE Pervasive Computing,* July-September 2007

[96] Building smart communities, online at http://www.holyroodconnect.com /tag/smart-cities/

[97] Using Big Data to Create Smart Cities, online at http://informationstrate gyrsm.wordpress.com/2013/10/12/using-big-data-to-create-smart-cities/

[98] N. Maisonneuve, M. Stevens, M. E. Niessen, L. Steels, "NoiseTube: Measuring and mapping noise pollution with mobile phones", in *Information Technologies in Environmental Engineering (ITEE 2009),* Proceedings of

the 4th International ICSC SymposiumThessaloniki, Greece, May 28–29, 2009

[99] J-S. Lee, B. Hoh, "Sell your experiences: a market mechanism based incentive for participatory sensing", *2010 IEEE International Conference onPervasive Computing and Communications (PerCom)*, pp.60–68, March 29, 2010, - April 2, 2010.

[100] R. Herring, A. Hofleitner, S. Amin, T. Nasr, A. Khalek, P. Abbeel, and A. Bayen, "Using Mobile Phones to Forecast Arterial Traffic Through Statistical Learning", *89th Transportation Research Board Annual Meeting*, Washington D.C., January 10–14, 2010

[101] M. Kranz, L. Roalter, and F. Michahelles, "Things That Twitter: Social Networks and the Internet of Things", in *What can the Internet of Things do for the Citizen (CIoT) Workshop* at *The Eighth International Conference on Pervasive Computing (Pervasive 2010)*, Helsinki, Finland, May 2010

[102] O. Vermesan, et al., "Internet of Energy – Connecting Energy Anywhere Anytime" in Advanced Microsystems for Automotive Applications 2011: Smart Systems for Electric, Safe and Networked Mobility, Springer, Berlin, 2011, ISBN 978–36-42213–80-9

[103] W. Colitti, K. Steenhaut, and N. De Caro, "Integrating Wireless Sensor Networks with the Web," Extending the Internet to Low Power and Lossy Networks (IP+ SN 2011), 2011 online at http://hinrg.cs.jhu.edu/joomla/images/stories/IPSN_2011_koliti.pdf

[104] M. M. Hassan, B. Song, and E. Huh, "A framework of sensor-cloud integration opportunities and challenges", in *Proceedings of the 3rd International Conference on Ubiquitous Information Management and Communication*, ICUIMC 2009, Suwon, Korea, January 15–16, pp. 618–626, 2009

[105] M. Yuriyama and T. Kushida, "Sensor-Cloud Infrastructure – Physical Sensor Management with Virtualized Sensors on Cloud Computing", NBiS 2010: 1–8

[106] C. Bizer, T. Heath, K. Idehen, and T. Berners-Lee, "Linked Data on the Web", *Proceedings of the 17th International Conference on World Wide Web (WWW'08)*, New York, NY, USA, ACM, pp.1265–1266, 2008

[107] T. Heath and C. Bizer, "Linked Data: Evolving the Web into a Global Data Space", *Synthesis Lectures on the Semantic Web: Theory and Technology*, 1st edition. Morgan & Claypool, 1:1, 1–136, 2011

[108] IBM, "An architectural blueprint for autonomic computing", IBM White paper. June 2005

[109] "Autonomic Computing: IBM's perspective on the state of Information Technology", 2001, http://www.research.ibm.com/autonomic/manifesto /autonomic_computing.pdf

[110] Connected Devices for Smarter Home Environments, IBM Data Magazine, 2014, online at http://ibmdatamag.com/2014/04/connected-devices-for-smarter-home-environments/

[111] International Conference on Autonomic Computinghttp://www. autonomic-conference.org/

[112] IEEE International Conferences on Self-Adaptive and Self-Organizing Systems, http://www.saso-conference.org/

[113] International Symposium on Software Engineering for Adaptive and Self-Managing Systems, http://www.seams2012.cs.uvic.ca/

[114] Awareness project, Self-Awareness in Autonomic Systemshttp://www .aware-project.eu/

[115] M. C. Huebscher, J. A. McCann, "A survey of autonomic computing – degrees, models, and applications", *ACM Computing Surveys (CSUR)*, Volume 40 Issue 3, August 2008

[116] A. S. Rao, M. P. Georgeff, "BDI Agents: From Theory to Practice", in *Proceedings of The First International Conference on Multi-agent Systems (ICMAS)*,1995. pp.312–319

[117] G. Dimitrakopoulos, P. Demestichas, W. Koenig, *Future Network & Mobile Summit 2010 Conference Proceedings.*

[118] John Naisbit and Patricia Aburdene (1991), *Megatrends 2000*, Avon

[119] D. C. Luckham, Event Processing for Business: Organizing the Real-Time Enterprise, John Wiley & Sons, 2012.

[120] T. Mitchell, *Machine Learning*, McGraw Hill, 1997

[121] D. Estrin, "Participatory Sensing: Applications and Architecture, online at http://research.cens.ucla.edu/people/estrin/resources/conferences/ 2010-Estrin-participatory-sensing-mobisys.pdf O. Etzion, P. Niblett, *Event Processing in Action*, Manning, 2011

[122] V. J. Hodgem, J. Austin, "A Survey of Outlier Detection Methodologies", *Artificial Intelligence Review*, 22 (2), pages 85–126, 2004.

[123] F. Angiulli, and C. Pizzuti, "Fast outlier detection in high dimensional spaces" in *Proc. European Conf. on Principles of Knowledge Discovery and Data Mining*, 2002

[124] H. Fan, O. Zaïane, A. Foss, and J. Wu, "Nonparametric outlier detection for efficiently discovering top-n outliers from engineering data", in *Proc. Pacific-Asia Conf. on Knowledge Discovery and Data Mining (PAKDD)*, Singapore, 2006

[125] A. Ghoting, S. Parthasarathy, and M. Otey, "Fast mining of distance-based outliers in high dimensional spaces", in *Proc. SIAM Int. Conf. on Data Mining (SDM)*, Bethesda, ML, 2006

[126] G. Box, G. Jenkins, *Time series analysis: forecasting and control*, rev. ed., Oakland, California: Holden-Day, 1976

[127] J. Hamilton, *Time Series Analysis*, Princeton Univ. Press, 1994

[128] J. Durbin and S.J. Koopman, *Time Series Analysis by State Space Methods*, Oxford University Press, 2001

[129] R. O. Duda, P. E. Hart, D. G. Stork, *Pattern Classification, 2nd Edition*, Wiley, 2000

[130] C.M. Bishop, *Neural Networks for Pattern Recognition*, Oxford University Press, 1995

[131] C. M. Bishop, *Pattern Recognition and Machine Learning*, Springer, 2006

[132] M. J. Zaki, "Generating non-redundant association rules", *Proceedings of the sixth ACM SIGKDD international conference on Knowledge discovery and data mining*, 34–43, 2000

[133] M. J. Zaki, M. Ogihara, "Theoretical foundations of association rules", *3rd ACM SIGMOD Workshop on Research Issues in Data Mining and Knowledge Discovery*, 1998

[134] N. Pasquier, Y. Bastide, R. Taouil, L. Lakhal, "Discovering Frequent Closed Itemsets for Association Rules", *Proceedings of the 7th International Conference on Database Theory*, (398–416), 1999

[135] C. M. Kuok, A. Fu, M. H. Wong, "Mining fuzzy association rules in databases", *SIGMOD Rec.* 27, 1 (March 1998), 41–46.

[136] T. Kohonen, *Self-Organizing Maps*, Springer, 2001

[137] S.-H. Hamed, S. Reza, "TASOM: A New Time Adaptive Self-Organizing Map", *IEEE Transactions on Systems, Man, and Cybernetics–Part B: Cybernetics* 33 (2): 271–282, 2003

[138] L.J.P. van der Maaten, G.E. Hinton, "Visualizing High-Dimensional Data Using t-SNE", *Journal of Machine Learning Research* 9(Nov): 2579–2605, 2008

[139] I. Guyon, S. Gunn, M. Nikravesh, and L. Zadeh (Eds), *Feature Extraction, Foundations and Applications*, Springer, 2006

[140] Y. Bengio, "Learning deep architectures for AI", *Foundations and Trends in Machine Learning*, 2(1):1–12, 2009

[141] Y. Bengio, Y. LeCun, "Scaling learning algorithms towards AI", *Large Scale Kernel Machines*, MIT Press, 2007

[142] B. Hammer, T. Villmann, "How to process uncertainty in machine learning?", *ESANN'2007 proceedings - European Symposium on Artificial Neural Networks*, Bruges (Belgium), 2007

[143] J. Quinonero-Candela, C. Rasmussen, F. Sinz, O. Bousquet, and B. Schölkopf, "Evaluating Predictive Uncertainty Challenge", in *Machine Learning Challenges: Evaluating Predictive Uncertainty, Visual Object Classification, and Recognising Tectual Entailment,* First PASCAL Machine Learning Challenges Workshop (MLCW 2005), Springer, Berlin, Germany, 1–27, 2006

[144] D. Koller and N. Friedman, *Probabilistic graphical models: principles and techniques*, MIT press, 2009

[145] M. R. Endsley, "Measurement of situation awareness in dynamic systems", *Human Factors*, 37, 65–84, 1995

[146] R. Fuller, *Neural Fuzzy System*, Åbo Akademi University, ESF Series A: 443, 1995, 249 pages. [ISBN 951–650-624–0, ISSN 0358–5654]

[147] S. Haykin, *Neural Networks: A Comprehensive Foundation, 2nd edn.*, Prentice-Hall, New York (1999)

[148] L. Rabiner, "A Tutorial on Hidden Markov Models and Selected Applications in Speech Recognition," *Proceedings of the IEEE*, vol. 77, no. 2, Feb. 1989

[149] S.K. Murthy, "Automatic construction of decision trees from data: a multi-disciplinary survey", *Data Mining Knowledge Discovery*, 1998.

[150] A. El Gamal, and Y-H Kim,"Network Information Theory", *Cambridge University Press*, 2011

[151] Z. Ma, "An Electronic Second Skin", in *Science*, vol. 333, 830–831 12 August, 2011

[152] Body Area Networks, IEEE 802.15 WPAN Task Group 6 (TG6), online athttp://www.ieee802.org/15/pub/TG6.html

[153] M. Debbah, "Mobile Flexible Networks: Research Agenda for the Next Decade", 2008, online at http://www.supelec.fr/d2ri/flexibleradio/pub/atc-debbah.pdf

[154] S. Venkatesan, "Limits on transmitted energy per bit in a cellular wireless access network", Private communication, Radio Access Domain, Bell Labs, New Jersey, USA

[155] GreenTouch Consortium, online at www.greentouch.org

[156] GreenTouch, "Annual Report 2010–2011", online at http://www.greentouch.org/uploads/documents/GreenTouch_2010–2011_Annual_Report.pdf

[157] G. Rittenhouse et al., "Understanding Power Consumption in Data Networks: A Systematic Approach", Eco. White paper, Alcatel-Lucent Bell Labs, Nov. 2009

[158] A. Gluhak, M. Hauswirth, S. Krco, N. Stojanovic, M. Bauer, R. Nielsen, S. Haller, N. Prasad, V. Reynolds, and O. Corcho, "An Architectural Blueprint for a Real-World Internet", in *The Future Internet - Future Internet Assembly 2011: Achievements and Technological Promises*, Lecture Notes in Computer Science, Vol. 6656, 1 st Edition, Chapter 3.3 Interaction Styles, 2011

[159] G. Grov, Al. Bundy, C. B. Jones, and A. Ireland, "The Al4FM approach for proof automation within formal methods", Submission to *Grand Challenges in Computing Research 2010*, UKCRC, online at http://www.ukcrc.org.uk/grand-challenge/gccr10-sub-20.cfm

[160] C.A.R. Hoare, "Communicating Sequential Processes", Prentice Hall International, 1985 + 2004, ISBN 0131532715, and http://www.usingcsp .com/

[161] R. Milner, "Communicating and Mobile Systems: The calculus", *Cambridge University Press*, 1999, ISBN 0–521-65869–1

[162] S. Haller and C. Magerkurth, "The Real-time Enterprise: IoT-enabled Business Processes", IETF IAB Workshop on Interconnecting Smart Objects with the Internet, March 2011

[163] OMG, Business Process Model and Notation specification, available at http://www.omg.org/technology/documents/br_pm_spec_catalog.htm, last accessed: November 15, 2011

[164] OASIS, Web Services Business Process Execution Language, http://docs.oasis-open.org/wsbpel/2.0/wsbpel-v2.0.html, last accessed: November 15, 2011.

[165] W3C, Unified Service Description Language Incubator Group, online at http://www.w3.org/2005/Incubator/usdl/, last accessed: November 15, 2011

[166] makeSense, EU FP7 Project, online at http://www.project-makesense.eu/, last accessed: November 15, 2011

[167] J. Hellerstein, "Parallel Programming in the Age of Big Data", 2008, online at http://gigaom.com/2008/11/09/mapreduce-leads-the-way-for-parallel-programming/

[168] E. Dans, " Big Data: a small introduction", 2011, Retrieved from online at http://www.enriquedans.com/2011/10/big-data-una-pequena-introduccion.html.

[169] A. Sheth, C. Henson, and S. Sahoo, "Semantic sensor web", *Internet Computing*, IEEE, vol. 12, no. 4, pp. 78–83, July-Aug. 2008.

[170] Open Geospatial Consortium, Geospatial and location standards,http://www.opengeospatial.org.

[171] M. Botts, G. Percivall, C. Reed, and J. Davidson, "oGC Sensor Web Enablement: Overview and High Level Architecture", *The Open Geospatial Consortium*, 2008, online at http://portal.opengeospatial.org/files/? artifact id=25562

[172] W3C Semantic Sensor Network Incubator Group, Incubator Activity, online at http://www.w3.org/2005/Incubator/ssn/

[173] Semantic Sensor Network Incubator Group, State of the Art Survey, http://www.w3.org/2005/Incubator/ssn/wiki/State_of_the_art_survey.

[174] S. Decker and M. Hauswirth, "Enabling networked knowledge", in *CIA '08: Proceedings of the 12th international workshop on Cooperative Information Agents XII*, Berlin, Heidelberg: Springer-Verlag, pp. 1–15, 2008

[175] P. Barnaghi, M. Presser, and K. Moessner, "Publishing Linked Sensor Data", in *Proceedings of the 3rd International Workshop on Semantic Sensor Networks (SSN)*, Organised in conjunction with the International Semantic Web Conference (ISWC) 2010, November 2010

[176] Logical Neighborhoods, Virtual Sensors and Actuators, online at http://logicalneighbor.sourceforge.net/vs.html

[177] K. M. Chandy and W. R. Schulte, "What is Event Driven Architecture (EDA) and Why Does it Matter?", 2007, online at http://complexevents.com/?p=212, (accessed on: 25.02.2008)

[178] D. Luckham, "What's the Difference Between ESP and CEP?", 2006, online at http://complexevents.com/?p=103, accessed on 15.12.2008

[179] The CEP Blog, http://www.thecepblog.com/

[180] EnOcean - the Energy Harvesting Wirless Standard for Building Automation and Industrial Automation, online at http://www.enocean.com/en/radio-technology/

[181] IEEE Std 802.15.4^{TM}-2006, Wireless Medium Access Control (MAC) and Physical Layer (PHY) Specifications for Low-Rate Wireless Personal Area Networks (LR-WPANs), online at http://www.ieee802.org/15/pub/TG4.html

[182] Bluetooth Low Energy (LE) Technology Info Site, online athttp://www.bluetooth.com/English/Products/Pages/low_energy.aspx

[183] The Official Bluetooth Technology Info Site, online at http://www.bluetooth.com/

[184] M-G. Di Benedetto and G. Giancola, *Understanding Ultra Wide Band Radio Fundamentals*, Prentice Hall, June 27, 2004

[185] ISO, International Organization for Standardization (ISO), Identification cards – Contactless integrated circuit(s) cards – Vicinity cards, ISO/IEC 14443, 2003

[186] N. Pletcher, S. Gambini, and J. Rabaey, "A 52 µW Wake-Up Receiver With 72 dBm Sensitivity Using an Uncertain-IF Architecture", in *IEEE Journal of Solid-State Circuits*, vol. 44, no1, January, pp. 269–280. 2009

[187] A. Vouilloz, M. Declercq, and C. Dehollain, "A Low-Power CMOS Super-Regenerative Receiver at 1 GHz", in *IEEE Journal of Solid-State Circuits*, vol. 36, no3, March, pp. 440–451, 2001

[188] J. Ryckaert, A. Geis, L. Bos, G. van der Plas, J. Craninckx, "A 6.1 GS/s 52.8 mW 43 dB DR 80 MHz Bandwidth 2.4 GHz RF Bandpass S-? ADC in 40 nm CMOS", in *IEEE Radio-Frequency Integrated Circuits Symposium*, 2010

[189] L. Lolis, C. Bernier, M. Pelissier, D. Dallet, and J.-B. Bégueret, "Band-pass Sampling RX System Design Issues and Architecture Comparison for Low Power RF Standards", *IEEE ISCAS 2010*

[190] D. Lachartre, "A 550 µW inductorless bandpass quantizer in 65 nm CMOS for 1.4-to-3 GHz digital RF receivers", *VLSI Circuits 2011*, pp. 166–167, 2011

[191] S. Boisseau and G. Despesse,"Energy Harvesting, Wireless Sensor Networks & Opportunities for Industrial Applications", in *EETimes*, 27th Feb 2012, online at http://www.eetimes.com

[192] J.G. Koomey, S. Berard, M. Sanchez, and H. Wong, "Implications of Historical Trends in the Electrical Efficiency of Computing", in *IEEE Annals of the History of Computing*, vol. 33, no. 3, pp. 46–54, March 2011

[193] eCall - eSafety Support, online at http://www.esafetysupport.org/en/ecall_toolbox/european_commission/index.html

[194] European Commission, "Smart Grid Mandate, Standardization Mandate to European Standardisation Organisations (ESOs) to support European Smart Grid deployments", M/490 EN,,Brussels 1st March, 2011

[195] Global Certification Forum, online at http://www.globalcertification forum.org

[196] SENSEI, EU FP7 project, online at http://www.sensei-project.eu

[197] IoT-A, EU FP7 project, online at http://www.iot-a.eu

[198] IoT6, EU FP7 project, online at http://www.iot6.eu

[199] IoT@Work, EU FP7 project, online at https://www.iot-at-work.eu/

[200] Federated Object Naming Service, GS1, online at http://www.gs1.org/gsmp/community/working_groups/gsmp#FONS

[201] Directive 2003/98/EC of the European Parliament and of the Council on the reuse of public sector information, 17 November 2003, online at http://ec.europa.eu/information_society/policy/psi/docs/pdfs/directive/psi_directive_en.pdf

[202] INSPIRE, EU FP7 project, – Infrastructure for Spatial Information in Europe, online at http://inspire.jrc.ec.europa.eu/

[203] H. van der Veer, A. Wiles, "Achiveing Technical Interoperability – the ETSI Approach", ETSI White Paper No.3, 3rd edition, April 2008, http://www.etsi.org/images/files/ETSIWhitePapers/IOP%20whitepaper%20Edition%203%20final.pdf

[204] Ambient Assisted Living Roadmap, AALIANCE

[205] Atmel AVR Xmega Micro Controllers, http://it.mouser.com/atmel_xmega/

[206] Worldwide Cellular M2M Modules Forecast, Beecham Research Ltd, August 2010

[207] Future Internet Assembly Research Roadmap, FIA Research Roadmap Working Group, May 2011

[208] D. Scholz-Reiter, M.-A. Isenberg, M. Teucke, H. Halfar, "An integrative approach on Autonomous Control and the Internet of Things", 2010

[209] NIEHS on EMF, http://www.niehs.nih.gov/health/topics/agents/emf/

[210] R.H. Weber/R. Weber, "Internet of Things - Legal Perspectives", Springer, Berlin 2010

[211] "The Global Wireless M2M Market", Berg Insight, 2010, http://www.berginsight.com/ReportPDF/ProductSheet/bi-gwm2m-ps.pdf

[212] M. Hatton, "Machine-to-Machine (M2M) communication in the Utilities Sector 2010–2020", Machina Research, July 2011.

[213] G. Masson, D. Morche, H. Jacquinot, and P. Vincent, "A 1 nJ/b 3.2–4.7 GHz UWB 50 Mpulses/s Double Quadrature Receiver for Communication and Localization", in *ESSCIRC 2010*

4

Internet of Things Global Standardisation - State of Play

Patrick Guillemin,[1] Friedbert Berens,[2] Marco Carugi,[3] Henri Barthel,[4] Alain Dechamps,[5] Richard Rees,[6] Carol Cosgrove-Sacks,[7] Jamie Clark,[8] Marilyn Arndt,[9] Latif Ladid,[10] George Percivall,[11] Bart De Lathouwer,[11] Steve Liang,[12] Ovidiu Vermesan,[13] Peter Friess,[14]

[1]*ETSI, France*
[2]*FB Consulting S.à r.l, Luxembourg*
[3]*ITU-T, Switzerland*
[4]*GS1 Global, Belgium*
[5]*CENELEC, Belgium*
[6]*CEN, Belgium*
[7]*OASIS, Switzerland*
[8]*OASIS, USA*
[9]*Orange, France*
[10]*University of Luxembourg, Luxembourg*
[11]*Open Geospatial Consortium, USA*
[12]*University of Calgary, Canada*
[13]*SINTEF, Norway*
[14]*European Commission, Belgium*

4.1 Introduction

The Information and Communication Technology development generates more and more things/objects that are becoming embedded with sensors and having the ability to communicate with other objects, that is transforming the physical world itself into an information and knowledge system.

Internet of Things (IoT) enables the things/objects in our environment to be active participants, i.e., they share information with other stakeholders or members of the network; wired/wireless, often using the same Internet Protocol (IP) that connects the Internet. In this way the things/objects are capable of recognizing events and changes in their surroundings and are acting and reacting autonomously largely without human intervention in an appropriate way.

The growth of interconnected things is expanding, and they use wireless and 2G/3G/4G mobile networks and 5G in the future.

The Internet of Things is bridging the virtual world with the physical world and the mobile networks need to scale to match the demands of 25–50 billion things. In this context it is needed to address the developments in the virtual world and the physical world in order to address the challenges of Internet of Things applications. In the virtual world, network virtualization, software defined networks, device management platforms, cloud computing and big data science are developing fast and need to be addressed as enabling technologies for Internet of Things.

In the physical world, the new wireless technologies for personal, home area networks, metropolitan and regional area networks all promise to deliver better economies of scale in terms of cost, energy and number of connections. Bringing the "Internet of Things" to life requires a comprehensive systems approach, inclusive of intelligent processing and sensing technology, connectivity, software and services, along with an ecosystem to address the smart environments applications.

The elements related to mobile networks, enabling scalability, large sensor (and actuator) networks, network virtualization, software defined networks, device management platforms, service oriented networks, cloud computing and big data to address the challenges related to standardisation.

These future IoT developments need to see acceleration and a maturing of common standards, more cross-sector collaboration and creative approaches to business models.

4.1.1 General

The Internet of Things (IoT) concept/paradigm is broad in its scope and the potential standards landscape is very large and complex. Technology is evolving and do not represent a barrier to adoption.

In the area of IoT, Europe is addressing the competitiveness in the context of globalisation. The technological specialisations built up over decades are transforming rapidly. In the area of IoT the IERC- Internet of Things European Research Cluster is focusing on increasing the link of projects, companies, organizations, people and knowledge at European level as a way of making projects more innovative and competitive.

Standards are needed for interoperability both within and between domains. Within a domain, standards can provide cost efficient realizations of solutions, and a domain here can mean even a specific organization or enterprise realizing an IoT. Between domains, the interoperability ensures cooperation between the engaged domains, and is more oriented towards a proper "Internet of Things". There is a need to consider the life-cycle process in which standardization is one activity. Significant attention is given to the "pre-selection" of standards through collaborative research, but focus should also be given to regulation, legislation, interoperability and certification as other activities in the same life-cycle. For IoT, this is of particular importance.

IERC is working to create a reference for pre-standardisation activities of EC IoT research projects that is the base for the position paper and the IoT standardisation roadmap. This effort has as goal to increase overall efficiency and raise mutual awareness, defragment and synergize in one unique place important information for stakeholders: Industry, Standard Development Organisations (SDOs), European Commission (EC).

A complexity with IoT comes from the fact that IoT intends to support a number of different applications covering a wide array of disciplines that are not part of the ICT domain. Requirements in these different disciplines can often come from legislation or regulatory activities. As a result, such policy making can have a direct requirement for supporting IoT standards to be developed. It would therefore be beneficial to develop a wider approach to standardization and include anticipation of emerging or on-going policy making in target application areas, and thus be prepared for its potential impact on IoT-related standardization. IoT implementation costs are expected to follow Moore's law. Targeting $1 chip sets by 2014, with a 15 year life for low bandwidth M2M apps such as smart meter reading. In this context standardisation has to be in place in order to gain full deployment potential.

A typical example is the standardization of vehicle emergency call services called eCall driven from the EC [5]. Based on the objective of increased road safety, directives were established that led to the standardization of solutions for services and communication by e.g. ETSI, and subsequently 3GPP. Another example is the Smart Grid standardization mandate M/490 [6] from the EC towards the European Standards Organisations (ESOs), and primarily ETSI, CEN and CENELEC.

The standardization bodies are addressing the issue of interoperable proto-col stacks and open standards for the IoT. This includes as well expanding the HTTP, TCP, IP stack to the IoT-specific protocol stack. This is quite challenging considering the different wireless protocols like ZigBee, RFID, Bluetooth, BACnet 802.15.4e, 6LoWPAN, RPL CoAP, AMQP and MQTT. Some of these protocols use different transport layers. HTTP relies on the Transmission Control Protocol (TCP). TCP's flow control mechanism is not appropriate for LLNs and its overhead is considered too high for short-lived transactions. In addition, TCP does not have multicast support and is rather sensitive to mobility. CoAP is built on top of the User Datagram Protocol (UDP) and therefore has significantly lower overhead and multicast support [8].

Any IoT related standardization must pay attention to how regulatory measures in a particular applied sector will eventually drive the need for standardized efforts in the IoT domain.

Agreed standards do not necessarily mean that the objective of interoper-ability is achieved. The mobile communications industry has been successful not only because of its global standards, but also because interoperability can be assured via the certification of mobile devices and organizations such as the Global Certification Forum [7] which is a joint partnership between mobile network operators, mobile handset manufacturers and test equipment manufacturers. Current corresponding M2M efforts are very domain specific and fragmented. The emerging IoT and M2M dependant industries should also benefit from ensuring interoperability of devices via activities such as conformance testing and certification on a broader scale.

To achieve this very important objective of a "certification" or validation programme, there is also a need of non-ambiguous test specifications which are also standards. This represents a critical step and an economic issue as this activity is resource consuming. As for any complex technology, implementation of test specifications into cost-effective test tools should also be considered. A good example is the complete approach of ETSI using a methodology (e.g. based on TTCN-3) considering all the needs for successful certification programmes.

The conclusion therefore is that just as the applied sector can benefit from standards supporting their particular regulated or mandated needs, equally, these sectors can benefit from conforming and certified solutions, protocols and devices. This is certain to help the IoT- supporting industrial players to succeed.

It is worth noting that setting standards for the purpose of interoperability is not only driven by proper SDOs, but for many industries and applied sectors it can also be driven by Special Interest Groups, Alliances and the Open Source communities. It is of equal importance from an IoT perspective to consider these different organizations when addressing the issue of standardization.

From the point of view of standardisation IoT is a global concept, and is based on the idea that anything can be connected at any time from any place to any network, by preserving the security, privacy and safety. The concept of connecting any object to the Internet could be one of the biggest standardization challenges and the success of the IoT is dependent on the development of interoperable global standards. In this context the IERC position is very clear.

Global standards are needed to achieve economy of scale and interworking. Wireless sensor networks, RFID, M2M are evolving to intelligent devices which need networking capabilities for a large number of applications and these technologies are "edge" drivers towards the "Internet of Things", while the network identifiable devices will have an impact on telecommunications networks. IERC is focussed to identify the requirements and specifications from industry and the needs of IoT standards in different domains and to harmonize the efforts, avoid the duplication of efforts and identify the standardization areas that need focus in the future.

To achieve these goals it is necessary to overview the international IoT standardization items and associated roadmap; to propose a harmonized European IoT standardisation roadmap; work to provide a global harmonization of IoT standardization activities; and develop a basic framework of standards (e.g., concept, terms, definition, relation with similar technologies).

The main issue today is how to organize, divide and prioritize the standardisation activities to focus on the aspects that provide the greatest customer benefit towards the goal of accelerating the rate of deployment and achieving interoperable and secure IoT applications.

Another main challenge is that IoT applications need to use standards developed separately by different groups or Technical Committees.

Finally, IoT applications interoperability (both communication and semantic) and the certification need to be addressed. Guidelines need to be developed, including mechanisms for interoperability enforcement and, where appropriate, leverage commercial certification activities.

4.2 IoT Vision

In the area of IoT, Europe is addressing the competitiveness in the context of globalisation. The technological specialisations built up over decades are transforming rapidly. In the area of IoT the IERC is focusing on increasing the link of projects, companies, organizations, people and knowledge at European level as a way of making projects more innovative and competitive.

IoT is a global paradigm and the standardisation issues have to be addressed in the global view. This liaison concept with the stakeholders and the SDOs working in the area of IoT is of paramount importance and will help strengthen promote exchange of ideas, solutions, results and validation of these among different standardisation activities.

Standards enable innovation, and are key for interoperability, may improve safety and security, are drivers for emergence of new markets, facilitate introduction of technologies (such as IoT), enhance competition and can help to „de-verticalize" industry by sharing and inter-operation of tools and technology, reducing the development and deployment costs for IoT applications.

Standardization is a complex process that needs to involve customers, suppliers and competitors and sometimes "competes" by different committees and standardization bodies, addressing separates domains, technologies, communities (vertical and horizontal fragmentation).

As presented in Figure 4.1 standardisation is a time-lagged and long-term process, usually fixes 1 – 5 years old state of the art rather than state of science and technology and can take up to three years to complete.

IoT is considered in the global context and in order to compete globally Europe has to use the enormous potential existing in the synergies among the standardisation activities in different organisations.

Figure 4.1 IEC Standardisation process cycle

This approach provides a more transparent, inclusive and competitive framework for efforts to strengthen European IoT research efforts and will allow easier the work on common standards.

4.2.1 IoT Drivers

Internet of Things gives semiconductor growth opportunity with possibilities for the billions of M2M (Machine-to-Machine) connected devices and the smart devices that provide the man-machine user experience.

Internet of Things offers opportunities to semiconductor and system companies as the implementation of applications occurs at multiple levels: object with individual IoT devices based on MCU's with network connectivity that sense and control. The value of the IoT is realized at the sensing/actuating, internetworking and at the solution level, where the big data from the IoT object is used in solving specific problems and creating new services.

The next generation of Smart Connected Homes, Smart Connected Vehicles or Internet of Vehicles, Internet of Energy, Smart Grids, Smart Manufacturing and Smart Health will enable new apps that use the IoT real time sourced big data while improving users' lives.

Wireless IoT connectivity range from cellular M2M modules from cellular providers, to specific chips for WiFi, Zigbee, 6LoWPAN (IPv6 over Low power Wireless Personal Area Networks), and BLE (Bluetooth Low Energy).

The IoT opportunities require aligning with the right standards and the involvement of broad technologies including SOC's, power, connectivity, software, Big Data this is a real challenge.

4.2.2 IoT Definition

The Internet of Things had until recently different means at different levels of abstractions through the value chain, from lower level semiconductor through the service providers.

The Internet of Things is a "global concept" and requires a common definition. Considering the wide background and required technologies, from sensing device, communication subsystem, data aggregation and pre-processing to the object instantiation and finally service provision, generating an unambiguous definition of the "Internet of Things" is non-trivial.

The IERC is actively involved in ITU-T Study Group 13, which leads the work of the International Telecommunications Union (ITU) on standards for next generation networks (NGN) and future networks and has been part of the ITU-T IoT-GSI (IoT Global Standards Initiative) team which has formulated the following definition [2]: *"**Internet of things (IoT):** A global infrastructure for the information society, enabling advanced services by interconnecting (physical and virtual) things based on existing and evolving interoperable information and communication technologies". NOTE 1 – Through the exploitation of identification, data capture, processing and communication capabilities, the IoT makes full use of things to offer services to all kinds of applications, whilst ensuring that security and privacy requirements are fulfilled.*

NOTE 2 – From a broader perspective, the IoT can be perceived as a vision with technological and societal implications.

The IERC definition [3] states that IoT is *"A dynamic global network infrastructure with self-configuring capabilities based on standard and inter-operable communication protocols where physical and virtual "things" have identities, physical attributes, and virtual personalities and use intelligent interfaces, and are seamlessly integrated into the information network".*

4.3 IoT Standardisation Landscape

This section gives an overview of the standardisation activities related to IoT within international standard organisations, including CEN/ISO, CENELEC/IEC, ETSI, IEEE, IETF, ITU-T, OASIS, OGC, and oneM2M.

4.3.1 CEN/ISO and CENELEC/IEC

4.3.1.1 CEN/CENELEC overview

As noted in the Introduction, the three European SDO's, CEN, CENELEC and ETSI play complementary roles in the development of IoT standards for Europe and in the liaisons they form with other SDOs and Industry SIGs across the globe.

The primary mission of CEN/CENELEC in relation to the IoT standardisation is to work in the applications zone that exists on the edge of the Internet of Things. The prime tasks are to:

- Integrate sensor data with existing (typically) barcode and RFID driven systems to enhance their performance and effectiveness
- Integrate object data concepts into existing standardised applications and ensure connectivity between "classic" data capture/storage/access systems and the Future Internet, particularly in the area of object discovery services

- Define the new application systems which ubiquitous sensor networks will enable.

The IoT is often seen from the perspective of M2M systems development. Whilst this may be appropriate in the early development of the IoT, a reality is that IoT applications are designed by people to provide new services to the citizen. The scale, abstraction and significance of such applications indicate that citizens will require that a governance framework is put in place to manage risk and resilience in such systems. CEN/CENELEC, together with their National Body members, increasingly introduces practical implementation of EC societal objectives into their standards output, in particular in the area of data protection and privacy.

Recent major EC standards programmes include EC Mandate responses M/436 "RFID" and M/490 "Smart Grids", together with project work on Smart Housing and Electro-Mobility.

4.3.1.2 CEN technical bodies

CEN's core business is the development of standards that meet the needs of the market.

Standardization is performed in a 'bottom-up' approach, thereby ensuring the market relevance of the resulting deliverables [36].

It is sometimes thought that standardization is no longer fit-for-purpose where new technology is concerned. The reality is that ESO Technical Specifications can be produced to the same timescale as industry SIGs may achieve.

The standardization activities of CEN are steered by the CEN Technical Board (BT), who has full responsibility for the execution of CEN's work programme.

Standards are prepared by Technical Committees (TCs). Each TC has its own field of operation (scope) within which a work programme of identified standards is developed and executed.

TCs work on the basis of national participation by the CEN Members, where delegates represent their respective national point of view. This principle allows the TCs to take balanced decisions that reflect a wide consensus. A TC may establish one or more sub-committees in the case of large programs of work.

The actual standards development is undertaken by working groups (WGs) where experts, appointed by the CEN Members but speaking in a personal capacity, come together and develop a draft that will become the future standard. This reflects an embedded principle of 'direct participation' in the standardization activities [37].

Work is also performed by CEN together with its sister organizations CEN-ELEC [52] and ETSI [53]. The ways and means of cooperation are laid down in the Internal Regulations Part 2 [54]. Where ETSI is also involved, the work follows the principles of the CEN-CENELEC-ETSI Basic Cooperation Agreement and the associated Modes of Cooperation [36].

The output of TC's may be:

- European Norms (EN) which are legal documents which, although not European Law in themselves, are able to be referred to by EU Regulations and Directives.
- Technical Specifications (TS) which allow rapid stabilisation of technologies and applications.
- Technical Reports (TR)

Additionally, CEN/CENELEC may create Workshops, which are particularly relevant in emerging or rapidly changing technologies that require fast completion of technical specifications or research projects.

The output of a Workshop is a CEN and/or CENELEC Workshop Agreements (CWAs).

It is perfectly possible to deliver TS, TR and CWA document in a year or less.

4.3.1.3 European standards

European Standards (EN) are documents that have been ratified by one of the 3 European Standards Organizations, CEN, CENELEC or ETSI. They are designed and created by all interested parties through a transparent, open and consensual process.

European Standards are a key component of the Single European Market. Though rather technical and unknown to the general public and media, they represent one of the most important issues for business. Although often perceived as boring and not particularly relevant to some organizations, managers or users, they are actually crucial in facilitating trade and hence have high visibility among manufacturers inside and outside the European territory. A standard represents a model specification, a technical solution against which a market can trade. It codifies best practice and is usually state of the art.

In essence, standards relate to products, services or systems. Now, however, standards are no longer created solely for technical reasons but have also become platforms to enable greater social inclusiveness and engagement with technology, as well as convergence and interoperability within a growing market across industries.

But the European Standard is something much more relevant than this. The CEN-CENELEC Internal Regulations, Part 2, states that the EN (European Standard) "carries with it the obligation to be implemented at national level by being given the status of a national standard and by withdrawal of any conflicting national standard".

The fact that European Standards must be transposed into a national standard in all member countries guarantees that a manufacturer has easier access to the market of all these European countries when applying European Standards. This applies whether the manufacturer is based in the CENELEC territory or not. Member countries must also withdraw any conflicting national standard: the EN prevails over any national standard [38].

4.3.1.4 Technical specifications

A Technical Specification (TS) is a normative document made available by CEN/CENELEC in at least one of the three official languages (English, German, French). A TS is established and approved by a technical body by a weighted vote of CEN/CENELEC national members. The Technical Specification is announced and made available at national level, but conflicting national standards may continue to exist. A Technical Specification is not permitted to conflict with an EN or HD (Harmonization Document). A TS is reviewed every 3 years at the latest. The maximum lifetime of a TS is 6 years.

Technical Specifications are established with a view to serving, for instance, the purpose of:

- Publishing aspects of a subject which may support the development and progress of the European market.
- Giving guidance to the market on or by specifications and related test methods.
- Providing specifications in experimental circumstances and/or evolving technologies.

TSs are not amended but replaced by a new edition with a new date of edition. However, Corrigenda are possible [39].

4.3.1.5 Technical reports

A Technical Report (TR) is an informative document made available by CEN/CENELEC in at least one of the official languages, established and approved by a technical body by simple majority vote of CEN/CENELEC national members. A Technical Report gives information on the technical content of standardization work.

Technical Reports may be established in cases when it is considered urgent or advisable to provide information to the CEN/CENELEC national members, the European Commission, the EFTA Secretariat or other governmental agencies or outside bodies, on the basis of collected data of a different kind from that which is normally published as an EN.

The decision to develop a TR can be taken by the Technical Board (BT), by a CEN/CENELEC Technical Committee (TC), a Technical Subcommittee (SC) or by a BTTF.

The CEN/CENELEC technical body preparing the draft TR (prTR) is also responsible for its approval. TRs are approved either in a CENELEC TC voting meeting or by a vote by correspondence of the CEN/CENELEC national members. If approved, the TR is made available unchanged to CCMC. TRs are not amended but replaced by a new edition with the same number and new date of edition. However, Corrigenda are possible.

No time limit is specified for the lifetime of TRs, but it is recommended that TRs are regularly reviewed by the responsible technical body to ensure that they remain valid [40].

4.3.1.6 CENELEC workshop agreements (CWA)

A CEN/CENELEC Workshop Agreement (CWA) is a document made available by CEN/CENELEC in at least one of the official languages (English, German, French). It is an agreement, developed and approved by a CEN/CENELEC Workshop and owned by CEN/CENELEC as a publication, which reflects the consensus of identified individuals and organizations responsible for its content. The Workshop Agreement is announced and possibly made available at national level. Conflicting national normative documents may continue to exist. Revision of a Workshop Agreement is possible.

A CWA shall not conflict with a European Standard (EN) and a Harmonization Document (HD). A CWA shall be withdrawn if the publication of an EN and HD brings the CWA into conflict with the EN and HD.

The CWA is valid for 3 years or until its transformation into another deliverable. After 3 years, the CCMC consults the former Workshop participants to see whether a renewal for a further 3 years is appropriate; if not, the CWA should be withdrawn [41].

4.3.1.7 CEN members

CEN's National Members are the National Standardization Bodies (NSBs) of the 28 European Union countries, the Former Yugoslav Republic of Macedonia, and Turkey plus three countries of the European Free Trade

Association (Iceland, Norway and Switzerland). There is one member per country.

A National Standardization Body is the one stop shop for all stakeholders and is the main focal point of access to the concerted system, which comprises regional (European) and international (ISO) standardization. It is the responsibility of the CEN National Members to implement European Standards as national standards.

The National Standardization Bodies distribute and sell the implemented European Standard and have to withdraw any conflicting national standards [42].

Details regarding this status are given in CEN/CENELEC Guide 20 - Guide on membership criteria of CEN and CENELEC [55].

4.3.1.8 CEN/TC 225

CEN/TC 225 "Automatic Identification Technologies" is tasked with the standardization of [43]:

- Data carriers for automatic identification and data capture
- The data element architecture
- The necessary test specifications and of technical features for the harmonization of cross-sector applications
- Establishment of an appropriate system of registration authorities, and of means to ensure the necessary maintenance of standards.

The work items of CEN/TC 225 are assigned as appropriate to Work Groups (WG), displayed in Table 4.1:

CEN/TC 225 Work Programme

At the time of writing, the CEN/TC 225 work programme consists of the following projects shown in Table 4.2.

Table 4.1 Work Groups in CEN/TC225 with appurtenant work items [44]

CEN/TC 225/WG 1	Optical Readable Media
CEN/TC 225/WG 3	Security and data structure
CEN/TC 225/WG 4	Automatic ID applications
CEN/TC 225/WG 5	RFID, RTLS and on board sensors
CEN/TC 225/WG 6	Internet of Things - Identification, Data Capture and Edge Technologies

4.3.1.9 CENELEC

CENELEC standards processes are very similar to those of CEN. CENELEC concentrates most of its work on 2 major types of deliverable:

- The European Standard (EN) and
- The Harmonization Document (HD)

Table 4.2   Projects in CEN/TC 225 work programme[45]

Project reference	Status	Initial Date
FprEN 16656 (WI=00225060) Information technology - Radio frequency identification for item management - RFID Emblem (ISO/IEC 29160:2012, modified)	Approved (Published on 2014-06-16)	2011-09-08
FprEN 16570 (WI=00225067) Information technology - Notification of RFID - The information sign and additional information to be provided by operators of RFID application systems	Approved (Published on 2014-07-09)	2012-03-20
FprCEN/TR 16669 (WI=00225068) Information technology - Device interface to support ISO/IEC 18000-3	Published	2012-05-09
FprCEN/TS 16685 (WI=00225069) Information technology - Notification of RFID - The information sign to be displayed in areas where RFID interrogators are deployed	Published	2012-05-09
FprCEN/TR 16670 (WI=00225070) Information technology - RFID threat and vulnerability analysis	Published	2012-05-09
prEN 1573 rev (WI=00225077) Bar coding - Multi industry transport label	Published	2013-01-28
FprCEN/TR 16671 (WI=00225072) Information technology - Authorisation of mobile phones when used as RFID interrogators	Published	2012-05-09

FprEN 16571 (WI=00225073) Information technology - RFID privacy impact assessment process	Approved (Published on 2014-06-25)	2012-05-09
FprCEN/TR 16672 (WI=00225074) Information technology - Privacy capability features of current RFID technologies	Published	2012-05-09
FprCEN/TR 16673 (WI=00225075) Information technology - RFID privacy impact assessment analysis for specific sectors	Published	2012-05-09
FprCEN/TR 16674 (WI=00225076) Information technology - Analysis of privacy impact assessment methodologies relevant to RFID	Published	2012-05-09
FprCEN/TR 16684 (WI=00225071) Information technology - Notification of RFID - Additional information to be provided by operators	Published	2012-05-09

These two documents are referred to commonly as "standards" and must be implemented in all CENELEC member countries, who must also withdraw any conflicting standard.

There are a few differences in the implementation process of EN's and HD's. Basically, the EN must be transposed as it is, not adding or deleting anything. The process for HD's is a bit more flexible. It is the technical content that must be transposed, no matter the wording or how many documents are made of it.

In addition to these two major deliverables, CENELEC also produces and approves Technical Specifications, Technical Reports and Workshop Agreements in a similar manner to CEN.

4.3.1.10 Smart grids: EC Mandate M/490

The European Standardization Organization (ESOs), i.e. CEN, CENELEC and ETSI, accepted the standardization Mandate M/490 on smart grid standardization.

The focal point addressing the ESO's response to M/490 was the CEN, CENELEC and ETSI Smart Grids Coordination Group (SG-CG), built around the membership of a previous JWG [47].

Smart Grid

A smart grid is an electricity network that can integrate in a cost efficient manner the behaviour and actions of all users connected to it - generators, consumers and those that do both - in order to ensure economically efficient, sustainable power system with low losses and high levels of quality and security of supply and safety (as per the definition given by the Expert Group 1 of the EU Commission Task Force for Smart Grids).

Smart Grids and Standardization

Standardization is a key issue for smart grids due to the involvement of many different sectors along the value chain - from the generation to the appliances in the households. Because the smart grid is broad in its scope, the potential standards landscape is also very large and complex [47].

In March 2011, the European Commission and EFTA issued the Smart Grid Mandate M/490. This was accepted by the three European Standards Organizations (ESOs), CEN, CENELEC and ETSI in June 2011.

M/490 requested CEN, CENELEC and ETSI to develop a framework to enable ESOs to perform continuous standard enhancement and development in the smart grid field [48]. M/490 highlighted the following key points:

- The need for speedy action
- The need to accommodate a huge number of stakeholders and
- To work in a context where many activities are international.

In order to perform the requested mandated work, the ESOs established in July 2011, together with the relevant stakeholders, the CEN-CENELEC-ETSI Smart Grid Coordination Group (SG-CG), being responsible for coordinating the ESOs reply to M/490 (successor of the JWG on standards for smart grids).

In 2012, the SG-CG Group focussed on the following mandated aspects:

- A Technical Reference Architecture
- A Set of Consistent Standards
- Sustainable Standardization Processes

The ESOs also investigated standards for information security and data privacy [47]. These reports and additional information on the Smart Grids standardization activities are available on the CEN-CENELEC website [56].

CENELEC Project Smarthouse

Further co-ordination is needed for transition to a common standardisation process of all communicating home equipment and associated services, leading to coherent sets of standards and specifications of interoperability between ICT services & applications, advanced electronic devices (products), commands and controls, and networks in homes of European citizens.

The objective of the CENELEC SmartHouse Roadmap project (supported by the European Commission and the European Free Trade Association) is to provide strategic direction and co-ordination for the standardisation activities of the ESOs (ETSI, CEN and CENELEC), together with other bodies that are active in this space, in order to reflect properly the growth of the SmartHouse and all the services, applications, systems and networks associated with it and to encourage future market growth. The intention is to identify all existing initiatives or standardisation works in the area and to co-ordinate actions so that to the greatest extent existing and future work should deliver interoperable solutions for any 'SmartHouse' service or application.

Smarthouse Roadmap

The project deliverable, the Roadmap, identifies what is already available from whatever standardisation source, what is being developed, what additional work would be needed and what is redundant or duplicated was developed.

The Roadmap consists of a matrix of standardisation activities and referenced work that are clearly prioritised and sorted with regard to prerequisite activities as well as identifying specific areas where appropriate stakeholders may co-operate to identify and carry out future co-ordinated standards work.

CLC/TC 205 "Home and Building Electronic Systems" took the results of the project on board prior to the possible transfer of the SmartHouse Roadmap Project to an existing or newly established dedicated coordination group [49].

Electric Vehicles

Standardization of electric vehicles is becoming an important issue. The need for clean energy and the support provided by smart-grids have led to new European policies that encourage the deployment of charging infrastructures for electrical vehicles.

There has been recent work internationally concerning charger and connector standards, but this work is not complete. We need to make sure that international standards meet European needs.

In 2010, CEN and CENELEC established a Focus Group on European Electro-Mobility. In October 2011, CEN/CENELEC delivered its response to European Commission Mandate M/468 (charging of electric vehicles).

The CEN/CENELEC report 'Standardization for road vehicles and associated infrastructure' defines the specific standardization requirements for European electro-mobility.

One of the main recommendations of the report was to establish a CEN-CENELEC eMobility Co-ordination Group with the aim to support coordination of standardization activities on Electro-Mobility [50].

4.3.1.11 ISO/IEC JTC 1/SWG 05 on the Internet of Things (IoT)

ISO/IEC JTC1 has established Special Working Group (SWG) 5. This group has the task of identifying the standardization gaps for the Internet of Things to allow JTC 1 to consider where work needs to be consolidated in the future in this arena.

The terms of reference of ISO/IEC JTC 1/SWG 5 are to:

- Identify market requirements and standardization gaps for IoT
- Encourage JTC 1 SCs and WGs to address the need for ISO/IEC standards for IoT
- Facilitate cooperation across JTC 1 entities
- Promote JTC 1 developed standards for IoT and encourage them to be recognized and utilized by industry and other standards setting organizations
- Facilitate the coordination of JTC 1 IoT activities with IEC, ISO, ITU, and other organizations that are developing standards for IoT
- Periodically report results and recommendations to ISO/IEC JTC 1/SWG 3 on Planning
- Provide a written report of activities and recommendations to JTC 1 in advance of each JTC 1 plenary
- Study IoT Reference Architectures/Frameworks and provide a study report. This study report should be written so it could be referenced in a possible JTC 1 New Work Item Proposal on IoT. The report shall be made available to JTC 1 no later than the 2014 JTC 1 Plenary.

The purpose of ISO/IEC JTC 1/SWG 5 is not to develop or publish IoT related standards, but to coordinate with ISO/IEC JTC 1 subcommittees, working groups, and special working groups and with other standards organizations to help better identify and convey the needs and gaps in the IoT world [51].

ISO/IEC JTC 1/SWG 5 is made up of four Ad Hoc groups, each of which carries out specific tasks in relation to IoT. The four Ad Hoc groups of ISO/IEC JTC 1/SWG 5 are:

Table 4.3 AD Hoc Groups of ISO/IEC JTC1/SWG 5

Ad Hoc Group	Working Area
Ad Hoc Group 1	Common understanding of IoT including IoT Mind Map and stakeholders
Ad Hoc Group 2	Identifying market requirements
Ad Hoc Group 3	Standardization gaps and roadmap for IoT
Ad Hoc Group 4	Study of IoT Reference Architectures/Frameworks

4.3.1.12 ISO/IEC JTC 1/WG 7 Sensor Networks

JTC1/WG7 is a standardization working group of the joint technical committee ISO/IEC JTC1 of the International Organization for (ISO) and the International Electrotechnical Commission (IEC), which develops and facilitates standards within the field of sensor networks.

The terms of reference for ISO/IEC JTC 1/WG 7 are [57]:

In the area of generic solutions for sensor networks, undertake standardization activities that support and can be applied to the technical work of all relevant JTC 1 entities and to other standards organizations. This includes activities in sensor networks such as the following:

- Standardization of terminology
- Development of a taxonomy
- Standardization of reference architectures
- Development of guidelines for interoperability
- Standardization of specific aspects of sensor networks

In the area of application-oriented sensor networks, identify gaps and overlaps that may impact standardization activities within the scope of JTC 1. Further, share this information with relevant entities within and outside of JTC 1. Unless better managed within another JTC 1 entity, this Working Group may pursue the following standardization activities as projects:

- Addressing the technology gaps within the scope of JTC 1 entities
- Exploiting technology opportunities where it is desirable to provide common approaches to the use of sensor networks across application domains
- Addressing emerging areas related to M2M and IoT

In order to foster communication and sharing of information between groups working in the field of sensor networks:

- Seek liaison relationships with all relevant SCs/WGs

- Seek liaison relationships with other organizations outside JTC 1
- Consider the possibility of conducting joint products with relevant ITU-T SGs
- Seek input from relevant research projects and consortia

ISO/IEC JTC 1/WG 7 currently has a number of standards published or under development within the field of sensor networks, including the ones shown in Table 4.4:

4.3.1.13 ISO/IEC JTC 1/SC 31 Automatic identification and data capture techniques

JTC1/SC31 is a standardization working group of the joint technical committee ISO/IEC JTC1 of the International Organization for (ISO) and the International Electrotechnical Commission (IEC), which develops and facilitates standards within the field of automatic identification technologies. These technologies include 1D and 2D barcodes, active and passive RFID for item identification and OCR.

Particular emphasis is currently being placed on devising structured security methods for RFID systems, including the use of cryptology. Data protection and authentication will be key requirements for the ubiquitous networks being proposed.

JTC1/SC31 has developed a large catalogue of barcode and RFID standards which underpin the existing AIDC applications which capture, store and provide access to data. Increasingly, these applications operate without human intervention, with both data capture and system response being automatic with private intranets providing the transport and discovery mechanisms. These systems already provide the edge of the emerging Future Internet.

A number of SC31 experts are members of SWG5. The results of that work are awaited before further IoT work items are created.

CEN/TC225 has liaison status with SC31, and from time to time ISO standards are adopted as ENs through the UAP process. A recent example has been ISO 29160 RFID Emblem.

4.3.2 ETSI

ETSI is a producer of globally applicable standards for ICT, including fixed, mobile, radio, converged, broadcast and Internet technologies. The Institute is at the forefront of emerging technologies. It is building close relationships with research bodies and addressing the technical issues that will drive the economy of the future and improve life for the next generation. For

Table 4.4 ISO/IEC JTC1/WG7 sensor networks standards [57]

ISO/IEC Standard	Title	Status	Description
ISO/IEC 29182-1	Information technology – Sensor networks: Sensor Network Reference Architecture (SNRA) – Part 1: General overview and requirements	Published (2013)	Provides a general overview of the characteristics of a sensor network and the organization of the entities that comprise such a network
ISO/IEC 29182-2	Information technology – Sensor networks: Sensor Network Reference Architecture (SNRA) – Part 2: Vocabulary and terminology	Published (2013)	Facilitates the development of International Standards in sensor networks by presenting terms and definitions for selected concepts relevant to the field of sensor networks
ISO/IEC 29182-3	Information Technology – Sensor Networks: Sensor Network Reference Architecture (SNRA) – Part 3: Reference architecture views	Under development	"Architecture views including business, operational, systems, and technical views which are presented in functional, logical, and/or physical where applicable"
ISO/IEC 29182-4	Information technology – Sensor networks: Sensor Network Reference Architecture (SNRA) – Part 4: Entity models	Published (2013)	Presents models for the entities that enable sensor network applications and services according to the SNRA
ISO/IEC 29182-5	Information technology – Sensor networks: Sensor Network Reference Architecture (SNRA) – Part 5: Interface definitions	Published (2013)	Provides the definitions and requirements of sensor network interfaces of the entities in the SNRA that covers the following aspects: • Interfaces between functional layers to provide service access for the modules in the upper layer to exchange messages with modules in the lower layer • Interfaces between entities introduced in the SNRA enabling sensor network services and applications

(Continued)

Table 4.4 Continued

ISO/IEC Standard	Title	Status	Description
ISO/IEC 29182-6	Information Technology – Sensor Networks: Sensor Network Reference Architecture (SNRA) – Part 6: Application Profiles	Under develop- ment	Describes • Functional blocks and components of a generic sensor network • Generic sensor network reference architecture incorporating the relevant sensor network-related base standards to support interoperability and data interchange
ISO/IEC 29182-7	Information Technology – Sensor Networks: Sensor Network Reference Architecture (SNRA) – Part 7: Interoperability guidelines	Under develop- ment	Provides • An overview of interoperability for heterogeneous sensor networks • Guidelines for interoperability between heterogeneous sensor networks
ISO/IEC 20005	Information technology – Sensor networks – Services and interfaces supporting collaborative information processing intelligent sensor networks	Published (2013)	Specifies services and interfaces supporting collaborative information processing (CIP) in intelligent sensor networks, which includes: • CIP functionalities and CIP functional model • Common services supporting CIP • Common service interfaces to CIP
ISO/IEC 30101	Information technology – Sensor Networks: Sensor Network and its interfaces for smart grid system	Under develop- ment	Describes: • Interfaces between the sensor networks and other networks • Sensor network architecture to support smart grid systems • Interface between sensor networks with smart grid systems

			• Sensor network based emerging applications and services to support smart grid systems
ISO/IEC 30128	Information technology – Sensor Networks – Generic Sensor Network Application Interface	Under development	Describes: • Generic sensor network applications' operational requirements • Sensor network capabilities • Mandatory and optional interfaces between the application layers of service providers and sensor network gateways

many years ETSI has been a driving force behind mobile communications systems, playing a prominent role as one of the founding partners of the Third Generation Partnership Project (3GPP). As well as long-established activities, ETSI is also responding to a new challenge in which ICT is either driving or facilitating other sectors such as transportation, utilities, eHealth, Cloud, smart cities, smart manufacturing and ambient assisted living.

The nature of ICT is pervasive and drives changes affecting the economy, society and politics. Society is increasingly dependent on ICT infrastructures which drive the needs for standards. In ETSI Long Term Strategy, ETSI aims to position itself closer to the research and innovation ecosystems of its members as well as European research and development programmes. ETSI maintains close links with the research community and participate in relevant European Commission Framework Programme 7 (FP7) and Horizon 2020 projects. In this way ETSI aims to identify new technologies with a standardisation need. ETSI role in these projects varies. For example, ETSI wide ranging expertise means it can help drive innovation in diverse areas, such as improving the quality of life through eHealth and Smart Personal Health (SPH). ETSI Centre for Testing and Interoperability (CTI) works on test specifications for a vehicle to grid interface for charging electric vehicles.

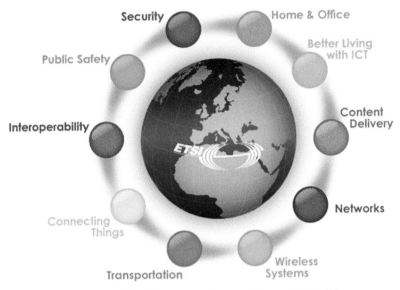

Figure 4.2   ETSI Technology Clusters (Source: ETSI [58])

ETSI is supporting the implementation and evolution of the European Union's Global Navigation Satellite System (GNSS) programmes, EGNOS and Galileo. The Institute is a partner in EC GSA Project SUNRISE [59], which runs the Open GNSS Service Interface Forum [60] for two industrial user groups of GNSS and future Galileo services, Location Based Services (LBS) and Intelligent Transport Systems (ITS). Links between LBS and the 'Internet of Things' (IoT) have been established. Here is an opportunity for GNSS, Augmented Reality and IoT to collaborate on LBS.

An ever increasing number of everyday machines and objects are now embedded with sensors or actuators and have the ability to communicate over the Internet. These 'smart' objects can sense and even influence the real world. Collectively they make up what is known as the 'Internet of Things'. The IoT draws together various technologies including Radio Frequency Identification (RFID), Wireless Sensor Networks (WSNs) and Machine-to-Machine (M2M) service platforms. ETSI is addressing the issues raised by connecting potentially billions of these 'smart objects' into a communications network, by developing the standards for data security, data management, data transport and data processing. This will ensure interoperable and cost-effective solutions, open up opportunities in new areas and allow the market to reach its full potential.

ETSI Machine-to-Machine Communications Technical Committee (TC M2M/TC smartM2M) is addressing the application independent 'horizontal' service platform within the M2M architecture which, with its evolved functionality, is capable of supporting a very wide range of services, including smart metering, smart grids, eHealth, city automation, consumer applications and car automation. ETSI will address the second phase of work in response to the European Commission (EC) mandate on Smart Metering (M/441), which includes security, use cases and the monitoring of deployments. ETSI responded to the 'Smart Grid Mandate' (M/490) with a discussion on a view to include the architectural models developed for M/441.

After oneM2M partnership project creation, in TC M2M, the future will be at System Level. New Terms of Reference have been endorsed by the ETSI board in September 2013 with Consolidation of transfer towards oneM2M and a rename of the TC M2M into TC SmartM2M. Considering the M2M platform environment at a system level with a focus on:

- Interoperability with M2M Area Network
- Interoperability with 3GPP Networks
- End to End Security and
- Introducing Additional functionalities and APIs (ex: Abstraction layer, data models)

ETSI TC SmartM2M will Publicize M2M platform (Provide tutorials and developers guides), Provide matched solutions in the framework of the 20-20-20 directive and standardization mandates on key issues: Security, Authentication /Identification and Interoperability. ETSI facilitates IoT semantic interoperability in TC SmartM2M.

As machine to machine applications proliferate, the market for embedded communications modules will expand. This introduces a challenge for both module manufacturers and those who integrate modules into applications to avoid fragmentation, improve volume and scalability, and facilitate integration and evolution. To address the need for new standards in this area ETSI has created an Industry Specification Group (ISG) for specifying a form factor for embedded modules to meet the requirements of emerging non-traditional mobile devices in support of embedded mobile services across multiple vertical markets. The standard will specify electrical as well as mechanical aspects, including the I/O interfaces, pad placement and module dimensions, targeting Surface Mount Technology (SMT) as manufacturing technology.

A large part of future IoT systems and devices will rely on wireless communication technologies deploying frequency spectrum as the main resource. In ETSI the Electromagnetic Compatibility and Radio Spectrum Matters Technical Committee (TC ERM) is responsible for the management and standardization of the spectrum usage for all wireless systems in Europe. In this sense TC ERM lays the ground for the future success of IoT by providing and managing the core resource for the communication.

TC ERM is maintaining all European Standard (EN) for spectrum usage. Some examples are given below:

- Radio LAN in 2.4GHz (ERM TG11)
- Cooperative ITS in 5.9GHz and 60GHz(ERM TG37)
- Short Range devices in 9KHz to 300GHz(ERM TG28)
- Ultra Wide Band in 3.1GHz to 100GHz(ERM TGUWB)
- RFID (ERM TG34)
- Wireless Industrial in 5.8GHZ (ERM TG41)

As part of these activities TC ERM actively identifies the spectrum requirements for existing and upcoming applications and drives the provision of the required spectrum resource by the European CEPT. This is done by providing CEPT with System Reference Documents (SRDoc) for all kinds of wireless applications like Smart Metering, Automotive, Home Automation, Smart Cities, Cooperative ITS and Alarms to guide spectrum allocations and regulation updated in Europe. The regulation results are used in TC ERM to generate Technical Specifications linking key market standards with European specific requirements and updated spectrum regulations. These harmonized standards are the basis for the market introduction of wireless devices based on the European Radio Equipment Directive.

As an important example for the work in TC ERM for the support of IoT the activities on Short Range Device (SRD) radio equipment in the 9 kHz - 300 GHz band should be mentioned. These activities are mainly managed in the TG28 (SRD), TG 34(RFID) and the TGUWB. In the scope of these activities the applications like RFIDs for national ID cards and passports, Near Field Communication (NFC) and also wireless charging are handled.

The main challenge for the future will be the identification and provisioning of the required spectrum resources for the upcoming IoT applications without scarifying existing applications.

Many of the connecting objects in M2M and the IoT need only low throughput connectivity. ETSI Industry Specification Group on Low Throughput Networks (ISG LTN) is specifying a new ultra low power network for very low data rates for ultra-long autonomy devices to provide an efficient connection that is both cost-effective and low in energy consumption.

ETSI ISG LTN has begun defining use cases and a dedicated architecture for LTN.

ETSI is also developing a new application of Digital Enhanced Cordless Telecommunications (DECT) Ultra Low Energy (ULE) for use in sensors, alarms, M2M applications and industrial automation.

ETSI Satellite Earth Stations and Systems Technical Committee develops standards for all types of terminals transmitting to a satellite: satellite terminals including hand portables, VSATs as well as devices mounted on aircraft, vessels, trains and vehicles. Recently a standard has been published for Earth station on board mobile platforms operating at Ka band.

CyberSecurity is providing the means for protecting the user and creating a more secure and profitable environment for industry and commerce. ETSI security work addresses numerous aspects including mobile/wireless communications, information technology infrastructure, lawful interception and data retention, electronic signatures, smart cards, fixed communications and security algorithms.

For ETSI, Security is a key element in standardisation and affects most areas of its work. ETSI Industry Specification Group on Information Security Indicators (ISG ISI) objectives are to address the full scope of main missing security event detection issues. Currently reference frameworks in the Cyber Defence and Security Information and Event Management (SIEM) fields are often missing or are still very poor, thus hindering IT security controls benchmarking. This is why we need IT security indicators which are related to event classification models.

ETSI is looking into the possible replacement of the Terrestrial Trunked Radio (TETRA) air interface encryption algorithm, and ETSI is developing a Technical Report on security-related use cases and threats in Reconfigurable Radio Systems (RRS). ETSI liaises with CEN as new standards are developed in response to EC Mandate 436 on the privacy and security of RFID. Other ongoing activities include Quantum Key Distribution. ETSI Industry Specification Group on Identity and access management for Networks and Services (ISG INS) is developing architecture and protocol specifications for advanced identity management in the Future Internet including the Internet of Things.

Improving the quality of health care, reducing medical costs and fostering independent living for those needing care are key objectives of the Digital Agenda for Europe. Telemedicine, for example, can improve the treatment of patients both at home and away, and reduces unnecessary hospitalisation. However, figures from the World Health Organisation show that only 8% of patients today use tele-monitoring. Medical issues are currently a key focus of work in TC ERM.

ETSI signed an important new co-operation agreement with its partner European Standards Organizations, CEN and CENELEC. ETSI has had an agreement in place since 1990 and has continuously co-operated since then. However, this new agreement will, for the first time, enable the creation of joint technical committees to produce joint standards which will be published by the three bodies. References: [15], [16], [17], [18], [19].

4.3.3 IEEE

4.3.3.1 Overview

It is predicted that 50 to 100 billion things will be electronically connected by the year 2020. This Internet of Things (IoT) will fuel technology innovation by creating the means for machines to communicate many different types of information with one another. With all objects in the world connected, lives will be transformed. But the success of IoT depends strongly on standardization, which provides interoperability, compatibility, reliability, and effective operations on a global scale. Recognizing the value of IoT to industry and the benefits this technology innovation brings to the public, the IEEE Standards Association (IEEE-SA) develops a number of standards, projects, and events that are directly related to creating the environment needed for a vibrant IoT [20].

The IEEE-SA has also recognized that the IoT incorporates aspects from many fields of technology. Here are some highlights from some of the fields encompassed in IoT.

The main focus of the IEEE standardisation activities are on the lower protocol layers namely the Physical layer and the MAC layer.

The IEEE laid an early foundation for the IoT with the IEEE802.15.4 standard for short range low power radios, typically operating in the industrial, scientific and medical (ISM) band. Having shown some limitations with the initial solutions such as ZigBee, the basic 15.4 MAC and PHY operations were enhanced in 2012 to accommodate the requirements of industrial automation and Smart Grid metering.

The new version of the standard introduced the 802.15.4g PHY, which allows for larger packets up to two Kilo-Octets and in particular comfortably fits the IPv6 minimum value for the maximum transmission unit (MTU) of 1280 octets, and the 802.15.4e MAC, which brings deterministic properties with the Time Slotted Channel Hopping (TSCH) mode of operation.

The value of the TSCH operation was initially demonstrated with the semi-proprietary wireless HART standard, which was further enhanced at the ISA as the ISA100.11a standard, sadly in an incompatible fashion.

IEEE ComSoc has appointed the key partners of the IoT6 project to lead the newly created IoT track within the Emerging Technologies Committee. IoT6 created a web site and attracted 400 members in the first 3 months: http://www.ipv6forum.com/iot//. IoT6 will use this platform to disseminate IoT6 solutions on a large scale basis. The Globecom IoT track is under preparation [4].

4.3.3.2 Cloud Computing

Cloud computing offers the promise of ubiquitous, scalable, on-demand computing resources provided as a service for everything from mobile devices to supercomputers. Cloud computing offers end consumers a "pay as you go" model— a powerful shift for computing towards a utility model like the electricity system, the telephone system, or more recently the Internet. IEEE is coordinating the support of cloud computing through its Cloud Computing Initiative, the first broad-based collaborative project for the cloud to be introduced by a global professional association.

The concept of a cloud operated by one service provider or enterprise interoperating with a cloud operated by another provider is a powerful means of increasing the value of cloud computing to industry and users. Such federation is called the "Intercloud." IEEE is creating technical standards for this interoperability. The IEEE Intercloud Testbed ("Testbed" for short) creates a global lab - to prove and improve the Intercloud, based on IEEE P2302 Draft Standard for Intercloud Interoperability and Federation. To that end, IEEE is partnering with companies, universities, and research institutions around the world to create a well-connected standards-based platform for the Intercloud. The IEEE Cloud Computing Testbed also could be used to experiment with other IEEE cloud computing products and services such as eLearning education modules [21].

4.3.3.3 eHealth

IEEE has many standards in the eHealth technology area, from body area networks to 3D modeling of medical data and personal health device communications. Another area is the IEEE 11073^{TM} family of standards, which is a group of standards under Health Informatics/Personal Health Device Communication, for data interoperability and architecture. IEEE 11073 standards are designed to help healthcare product vendors and integrators create devices and systems for disease management, health and fitness and independent living that can help save lives and improve quality of life for people worldwide. IEEE is part of a larger ecosystem and has active collaborative relationships with other global organizations such as:

- Health Level Seven International (HL7), with a focus on data exchange/delivery)
- Integrating the Healthcare Enterprise (IHE), with a focus on development domain integration and content profiles
- International Health Terminology Standards Development Organisation (IHTSDO) with a focus on Systematized Nomenclature of Medicine Clinical Terms (SNOMED) Clinical Terminology
- ISO and CEN, both of which adopt many of the IEEE 11073 standards

This allows IEEE standards to be developed and used within a framework for interoperable medical device communications worldwide. The growing IEEE 11073 family of standards is intended to support interoperable communications for personal health devices and convey far-ranging potential benefits, such as reducing clinical decision-making from days to minutes, reducing gaps and errors across the spectrum of healthcare delivery, and helping to expand the potential market for the medical devices themselves [22].

4.3.3.4 eLearning

The IEEE Learning Technology Standards Committee (LTSC) is chartered by the IEEE Computer Society Standards Activity Board to develop globally recognized technical standards, recommended practices, and guides for learning technology. The IEEE LTSC coordinates with other organizations, both formally and informally, that produce specifications and standards for learning technologies. The IEEE LTSC has activities in several eLearning areas, including Digital Rights Expression Languages, Computer Managed Instruction, Learning Object Metadata, and Resource Aggregation Models for Learning, Education and Training, Competency Data Standards [23].

4.3.3.5 Intelligent Transportation Systems (ITS)

IEEE has standards activities on many aspects of ITS, such as vehicle communications and networking (IEEE 802 series), vehicle to grid interconnectivity (IEEE P2030.1), addressing applications for electric- sourced vehicles and related support infrastructure, and communication for charging (IEEE 1901). In addition, the IEEE 1609 family of standards for Wireless Access in Vehicular Environments (WAVE) defines an architecture and a complementary, standardized set of services and interfaces that collectively enable secure vehicle-to-vehicle (V2V) and vehicle-to-infrastructure (V2I) wireless communications. Together these standards are designed to provide the foundation for a broad range of applications in the transportation environment, including vehicle safety, automated tolling, enhanced navigation, traffic management, and many others. As part of the global technology ecosystem, IEEE VTS/ITS collaborates and coordinates with many other organizations. IEEE VTS/ITS 1609 WG experts have participated in the exchange of IEEE draft documents to facilitate the expeditious development of profiles for use of IEEE drafts for European Norms (ENs) [24].

4.3.3.6 Network and Information Security (NIS)

IEEE has standardization activities in the network and information security space, including in the encryption, fixed and removable storage, and hard copy devices areas, as well as applications of these technologies in smart grids. IEEE's largest technical society, the IEEE Computer Society, is well equipped to provide technical expertise in network and information security efforts. For over thirty years, the IEEE Computer Society has had a technical committee focused on computer security and privacy. It publishes the well-respected IEEE Security & Privacy magazine, which offers articles by top thinkers in the information security industry, and sponsors two long-established premier technical meetings, the IEEE Security and Privacy Symposium and the Computer Security Foundation Workshop. IEEE-SA's Industry Connections Security Group is another important activity, providing a flexible and nimble platform for stakeholders to respond to the new malware environment. It has three key activities [25]:

- The Malware MetaData Exchange Format (MMDEF) Working Group, which develops the MMDEF format. It is used primarily by anti-virus companies and researchers to exchange information about malware and known clean files.
- The Stop-eCrime Working Group, which develops various resources (taxonomies, protocols, guidelines, etc.) to help stop electronic crime.
- The Privilege Management Protocols Working Group, which develops new mechanisms and protocols for efficient authentication and secure determination of "who" can do "what" in applications.

4.3.3.7 Smart Grid

IEEE-SA has many smart grid standards and projects in development from the diverse fields of digital information and controls technology, networking, security, reliability, assessment, interconnection of distributed resources including renewable energy sources to the grid, sensors, electric metering, broadband over power line, and systems engineering.

IEEE has established a wide range of relationships across many geographic and SDO boundaries. Coordination and collaboration across the standards community are necessary to ensure that the smart grid can realize its full potential. IEEE has relationships with organizations that allow partnership in the development of standards. Our partners in international collaboration include:

- International Electrotechnical Commission (IEC)
- International Organization for Standardization (ISO)
- International Telecommunication Union (ITU)
- Korean Agency for Technology and Standards (KATS)
- Korea Electronics Association (KEA)
- Korean Society of Automotive Engineers (KSAE)
- State Grid Corporation of China (SGCC)
- Telecommunication Technology Committee (TTC)
- Telecommunications Technology Association (TTA)
- European Telecommunications Standards Institute (ETSI)

IEEE also participates in groups such as the India Smart Grid Forum, the Smart Grid Interoperability Panel (SGIP), and the steering committee of the European Technology Platform for the Electricity Network of the Future (ETP SmartGrids).

The IEEE 2030 series for smart grid interoperability currently consists of half a dozen standards and ongoing projects. IEEE 2030 is based on a smart grid interoperability reference model (SGIRM) and provides alternative approaches and best practices for smart grid work worldwide. Also IEEE has been involved with smart grid technologies for many years, including integrating distributed resources that incorporate renewable energy (IEEE 1547). In addition, the IEEE 1547 series of standards deals with many different facets of renewable energy, such as microgrids (IEEE 1547.4) and secondary networks for distributed resources (IEEE 1574.6).

As the smart grid evolves, IEEE is looking at the next phase of the evolution with standards projects such as time synchronization (IEEE C37.238) and cyber security (IEEE PC37.240 and IEEE P1686) [26].

4.3.4 IETF

The most recognizable enhancement by ISA100.11a is probably the support of IPv6, which came with the 6LoWPAN Header Compression, as defined by the IETF.

Another competing protocol, WIAPA, was developed in parallel in China, adding to fragmentation of the industrial wireless automation market, and ultimately impeding its promised rapid growth.

A strong request is now coming from the early adopters, in the industrial Process Control space, for a single protocol that will unify those existing protocols in a backward compatible fashion, and extend them for distributed routing operations. Distributed operations are expected to lower the deployment costs and scale to thousands of nodes per wireless mesh network, enabling new applications in large scale monitoring. The 6TiSCH Working Group is being formed at the IETF to address the networking piece of that unifying standard.

Based on open standards, 6TiSCH will provide a complete suite of layer 3 and 4 protocols for distributed and centralized routing operation as well as deterministic packet switching over the IEEE802.15.4e TSCH MAC. Most of the required 6TiSCH components already exist at the IETF in one form or another and mostly require adaptation to the particular case, and 6TiSCH will mostly produce an architecture that binds those components together, and provide the missing glue and blocks either as in-house RFCs, or by pushing the work to the relevant Working Groups at the IETF.

Yet, there is at least one entirely new component required. That component, 6TUS, sits below the 6LoWPAN HC layer in order to place the frames on the appropriate time slots that the MAC supports, and switch frames that are propagated along tracks that represent a predetermined sequence of time slots along a path. Centralized routing is probably a case where work will be pushed outside of the 6TiSCHWG. That component will probably leverage work that was done at the Path Computation Element (PCE) Working Group, and require additions and changes such as operation over the CoAP protocol, and new methods for advertising links and metrics to the PCE. All this work probably belongs to the PCE WG. Another example is the adaptation of the IPv6 Neighbour Discovery (ND) protocol for wireless devices (WiND) that will extend the 6LoWPAN ND operation and will probably be conducted at the 6MAN working group in charge of IPv6 maintenance.

Distributed route computation and associated track reservation, on the other hand, can probably be addressed within the 6TiSCH Working

Group, as it is expected to trivially extend the existing RSVP and RPL protocols.

Same goes for PANA that may be extended to scale the authentication to the thousands of devices. The next step for this work is a so called BoF in July 2013 in Berlin. The BoF will decide whether a WG should be formed and determine the charter for that WG [4].

In November 2013 a new IETF WG, 6lo, was created. 6lo (IPv6 over Network of Resource Constrained Nodes) will continue the work of 6LoWPAN WG on IPv6-over-foo adaptation layer specifications. These specifications are based on the 6LoWPAN specifications RFC 4944, RFC 6282, and RFC 6775, but will not embody routing, which is out of scope. The workgroup will work closely with 6man (IPv6 Maintenance), intarea (Internet Area Working Group), lwig (Light-Weight Implementation Guidance), Core, and Roll. The WG drafts, as of January 2014 are:

- draft-ietf-6lo-btle-00
- draft-ietf-6lo-ghc-00
- draft-ietf-6lo-lowpan-mib-00
- draft-ietf-6lo-lowpanz-01

4.3.5 ITU-T

The Telecommunication Standardization Sector of the International Telecommunication Union (ITU-T) is progressing standardization activities on Internet of Things (IoT) since 2005.

After a report on "The Internet of Things", published by the ITU in 2005, the ITU-T established a Joint Coordination Activity (JCA-NID), which aimed at sharing information and performing coordination in the field of network aspects of Identification systems, including RFID. The JCA-NID supported the work of the ITU-T Study Groups which led to the approval of initial Recommendations in the areas of tag-based identification services, Ubiquitous Sensor Networks (USN) and Ubiquitous Networking, and their application in Next Generation Networks (NGN) environment.

With the official recognition in 2011 of the centrality of IoT in the evolution of future network and service infrastructures, the JCA-NID was renamed as JCA-IoT (Joint Coordination Activity on Internet of Things [61]) and the working structure of the IoT-GSI (IoT Global Standards Initiative [1]) was formally established.

Since then, the ITU-T activities related to IoT have greatly expanded and produced additional Recommendations spanning various areas of application (e.g. networked vehicles, home networks, mobile payments, e-health, machine oriented communications, sensor control networks, gateway applications, ubiquitous applications (u - plant farming etc.), energy saving in home networks), as well as IoT framework and transversal aspects (basic concepts and terminology, IoT common requirements, ecosystem and business models, web of things, IoT security and testing etc.) [4].

Beyond the above mentioned IoT focused activities and a number of IoT work items currently under development in the IoT-GSI within various ITU-T study groups (the main ones being SG11, SG13, SG16 and SG17), other potential future IoT studies are included in the "IoT-GSI work plan" (a living list of potential studies maintained by the IoT-GSI).

In addition, it has to be noted that there have been and are other ITU-T on-going studies closely related to the IoT. It is worthwhile to mention here:

- an effort on transversal aspects (the Focus Group on M2M Service Layer - see below)

- some efforts focused on specific IoT application domains, including the Focus Group (FG) on Smart Sustainable Cities, the collaboration initiative on Intelligent Transport Systems (ITS) communication standards, the FG on Car Communication (concluded in 2013), the FG on Smart Grid (concluded in 2011)

- ongoing studies with an indirect relationship with the IoT, related to Future Networks, Service Delivery Platforms and Cloud Computing

In parallel with the JCA-IoT's coordination efforts with external entities and its maintenance of a cross-SDO list of IoT standard specifications and associated roadmap (the "IoT Standards Roadmap", freely available from the JCA-IoT web page [62]), a remarkable milestone has been achieved by the IoT GSI via the finalization in June 2012 of the ITU-T Recommendation Y.2060 [2]. This Recommendation includes, among others, a definition of the IoT which has obtained large acceptance within the IoT community, including across different standards development organizations. It has to be noted , in this perspective, that the Machine to Machine (M2M) communication capabilities are seen as an essential enabler of the IoT, but represent only a subset of the whole set of capabilities of the IoT.

Among the various ITU-T IoT-related efforts, the Focus Group on M2M Service Layer (FG M2M) [9] deserves a special mention: established in 2012 with the key goal to study requirements and specifications for a common M2M Service Layer, it has focused its developments – from the point of view of use cases and derived requirements for the common M2M service layer - on the "e-health" application domain (specifically, on remote patient monitoring and assisted living services). The FG M2M, who had targeted the inclusion of vertical market stakeholders not part of the traditional ITU-T membership, such as the World Health Organization (WHO), and the collaboration with M2M and e-health communities and SDOs, has actually liaised with various SDOs, fora and consortia, including for the completion of an e-health standards repository.

The FG M2M work, whose last physical meeting was held in London in December 2013 and whose final electronic meeting on editorial aspects has taken place at the end of March 2014, has completed five deliverables [9] dealing with, respectively, e-health use cases, e-health ecosystem, M2M service layer requirements and architectural framework, overview of M2M service layer APIs and protocols, and e-health standards repository and gap analysis.

In the context of the FG M2M service layer work, in line with the IoT Reference Model described in ITU-T Y.2060, the M2M service layer capabilities aim to include those common to the support of different application domains as well as the specific ones required for the support of each application domain [4].

Lastly, it is worthwhile to mention the recent ITU-T workshop "IoT – Trends and Challenges in Standardization", Geneva, 18 February 2014, where the main achievements and current activities of ITU-T on IoT have been presented (together with inputs from other IoT standards related efforts, academic and open source communities) [10]:

IERC and ITU-T have entertained good relationships all along the IoT standardization activities of ITU-T, particularly in the context of JCA-IoT and IoT-GSI. IERC has liaised with ITU-T and taken an active role in the discussions which led to the finalization of the ITU-T definition of "Internet of Things" and the approval of ITU-T Y.2060 (aspects related to IoT Reference Model, IoT Ecosystem, high-level requirements of IoT and other IoT definitions). More recently, representatives of the IERC IoT-A project have actively participated and contributed to the progress of the ITU-T studies on IoT common requirements (ITU-T Y.2066) and, in perspective, their participation is expected on the progress of the ITU-T studies on IoT capabilities and functional architecture. On the other side, representatives of ITU-T have attended over the last period some IERC meetings in order to provide updates on ongoing ITU-T work and contribute to strengthen the international collaboration on IoT standards. The ITU and IERC collaboration and coordination are expected to continue in the future: beyond the ongoing collaboration on IoT framework aspects (including that expected on semantics and big data matters), it might involve also IoT "vertical" matters, for example e-health (ITU-T SG13 and SG16, whose completed work includes, respectively, ITU-T Y.2065 and H.810), Smart Cities (FG on Smart Sustainable Cities etc.), Intelligent Transport Systems (collaboration initiative on ITS communication standards) etc.

The IoT6 and IoT Lab projects will also maintain an on-going communication with the ITU-T through their Coordinator, Mandat International, which is a member of the ITU-T, following the JCA-IoT activities.

4.3.6 OASIS

Widely distributed networks of heterogeneous devices and sensors, as expected in the growing Internet of Things, require the agile combination of several advanced ICT methodologies, deployed together in massively scalable ways. Among other things:

- Network communications must use established basic patterns for *reliable transactional messaging and interaction*, and data protocols suitable for high-speed, high-volume transacting using vendor-neutral systems.
- Functions, services and actions must be made *modular and re-useable*, so that they can be shared and invoked by wide variety of different systems. Computing operations must be capable of being conducted across remote, distributed and parallel resources, to obtain the increased speed, easy scalability and ready availability available from *cloud computing* methodologies.
- The large volumes of data generated by these systems, which pervasively touch personal lives, businesses and locations, must have powerful and discrete *access control and cybersecurity capabilities*, so as to conform to public policy and business requirements for privacy and security.

4.3.6.1 Transactional Reliability

The basic requirements for reliable automated interaction patterns – covering a variety of logical needs such as resolution of duplicates, acknowledgements, the handling of sequentially ordered steps and Quality-of-Service – emerged over time as we learned how to design open, heterogeneous networked systems. Many of them were refined as part of a series of early standards serving the evolution of open middleware systems, such as W3C's Simple Object Access Protocol (SOAP) and OASIS's WS-Reliable Messaging and ebXML Messaging.

Today, as we create much larger and more loosely-coordinated systems, using cloud computing at scale – connecting huge numbers of often-computationally-low-powered devices and entities – our systems require more compact and simple transactional protocols that still support those logical needs, with a very low profile of resource use, but that also still hold up under ultra-high-volume and ultra-high-speed conditions. OASIS standards projects to fulfill that requirement include the OASIS **Message Queuing Telemetry Transport (MQTT) TC,** explicitly designed for IoT networks and based on the already-industry-deployed MQTT v3.1 and the Eclipse Foundation open source framework; and the OASIS standard **Advanced Message Queuing Protocol (AMQP),** widely used in the financial industry. Each project is developing a suite of related protocols and extensions for interoperability.

4.3.6.2 Modularity, reusability, and devices in the cloud

In created automated networks, one design choice that has persisted from the earliest days of e-commerce and web services is the need to encapsulate computing functions into re-usable and vendor-neutral services, so that they can be deployed in combinations, freely across systems and owners, like LEGO blocks that will snap together readily in multiple combinations. Early work to define and assure that outcome included OASIS's SOA Reference Model.

As the use of distributed, remote computing resources to build systems ("cloud computing") became widespread, individual services have been pressured to create endpoints which were widely useable by strangers; one method is by issuing defined instructions and calls for those functions that can be used by coders and systems, such as Application Programming Interfaces (APIs). As the degree of automation, and the number of services, has grown exponentially, the industry necessarily has developed more and more advanced methods for finding, running and coordinating those services, singly and in aggregations. IoT device networks have all those same cooperation and interoperability requirements – but often pushed down into much less computationally-robust devices. The "things" in those networks, and the protocols that employ and drive them, may have much more need for substitution, duplication and fail-over, when random bits of far-flung, barely-smart-devices fail, or respond only intermittently.

Industry standards projects to make that possible include the OASIS **Cloud Application Management for Platforms (CAMP) TC,** an interoperable protocol for self-service provisioning, monitoring, and control of portable applications, and the OASIS standard **Topology and Orchestration Specification for Cloud Applications (TOSCA),** which can be used to describe and direct cloud infrastructure services and applications across multiple networks and different providers.

OASIS members also have developed a set of open standards web services tools for device discovery and management, including the OASIS standards **Devices Profile for Web Services** and **Web Services Dynamic Discovery (WS-Discovery).** The OASIS **Open Services for Lifecycle Collaboration (OSLC)** projects apply the W3C's Linked Data Platform semantic methodology to describe, find parts of, and help control networks of far-flung networked devices and systems, with specific application to M2M and smart devices being addressed by the OASIS **OSLC Lifecycle Integration for TC.**

Another OASIS project, the **OASIS Biometrics TC,** is defining lightweight REST protocols for biometric security sensors and controls based on US NIST's WS-BD. Finally, a suite of OASIS standards projects for remote device interaction and control, developed in cooperation with and endorsed by the US Smart Grid Interoperability Program, provide open service scheduling, date and time, price and demand functions – permitting dynamic two-way interactions and queries, all the way from consumer home devices to regional utility infrastructure nodes and servers – via the **OASIS WS-Calendar** and **Energy Market Information Exchange TCs**.

4.3.6.3 All that big data from all those things: access control, cybersecurity and privacy

The sheer amount of data generated by far-flung device and sensor networks, when multiple systems are capable of being connected or jointly queried, is unprecedented in human history ... as are the privacy intrusion and security risk problems that it creates. When our technology makes it possible for external and even anonymous queries to reach sensors and servocontrols in every home, every business, and perhaps every shirt pocket in the world, our industry's policy and cybersecurity challenges are radically multiplied. This makes careful, pervasive application of deliberate methodologies for access control, rules implementations, and security on the wire essential. Privacy and security cannot be left behind, in the changing architecture of mobile and remote devices.

4.3.6.4 Access control

OASIS is the home of several of the most successful and widely-deployed open standards for access control and secure multiparty transacting. Our Security Assertion Markup Language (SAML) has been the most-widely-used and - known open standard for identity management for years, and most newer standards projects either use or duplicate its logical structure for assertions and authorization. It is widely tooled, and among other things is widely deployed in government and academic systems; SAML even drives authorization for ISO's own standards creation and document management platform. The OASIS eXtensible Access Control ML (XACML) provides advanced discrete access control capabilities including profiles for role-based access control, REST architecture, export controls and intellectual property license control.

Deploying that advanced functionality in wider networks with sparse control structures has been the subject of several advanced OASIS projects, including the OASIS Identity in the Cloud TC, whose gap analysis work has been widely used by global standards bodies to identify areas for additional standardization and the OASIS Cloud Authorization TC.

4.3.6.5 Encryption and cybersecurity

Standards bodies have produced a number of secure solutions for human-interfaced systems (such as Web sessions), going back all the way to the ITU's X.509 PKI certificates, and their use in the widely-deployed Secure Socket Layer (SSL) and IETF's HTTPS (RFC 2818). Standardized methods apply cybersecurity and authentication functions on the wire compactly in networked exchanges are a more recent development. OASIS projects that fulfil that demand include the OASIS standard Key Management Interoperability Protocol (KMIP), and the mobile and cloud computing functionality being added to the widely-used Public-Key Cryptography Standard #11 cryptography specification by the OASIS PKCS #11 TC. Both were recently demo'ed as key M2M and IoT cybersecurity tools at the 2014 RSA Conference. OASIS members also developed the OASIS standard SOAP-over-UDP, extending W3C's SOAP for use over the widely-used IETF RFC 768 User Datagram Protocol (UDP), a terse core Internet data transport method for simple systems.

4.3.6.6 Privacy

Much of the promise of the Internet of Things will be lost, if we cannot keep our promises both about functionality and the appropriate use of data. In addition to ensuring correct targeting of recipients (access control) and safety of data on the wire (e.g., encryption), systems *must* be able to accurately express and execute rules about the privacy of and limitations for the data that is exchanged. In highly-automated and networked transactions, it is essential that the legal and business rules for private and limited-access information be baked into the design of systems at the start; controls or interdictions applied afterwards as a last-minute thought have a long history of failure.

Key open standards projects for ensuring that privacy functions are native in networked systems include the OASIS **Privacy Management Reference Model (PMRM) TC,** which defines an openly-available privacy technical model and a structured, modeled set of common implementable services and interactions which can tie network functions and events to the fulfilment of policy requirements in auditable ways; and the OASIS **Privacy by Design for Software Engineers (PbD-SE) TC,** which is developing a privacy governance model for code expressed in, among other things, guidance for interface design and code tools including in OMG's UML.

4.3.7 OGC

The phrase "spatial is important" is almost always relevant to IoT. From a geographical information perspective, some important facts are: every sensor has a location, and location is always important. Secondly, outputs from multiple Internet-connected sensors sampling the same phenomena, such as temperature or salinity, can be aggregated to form a GIS data layer. Finally, IoT is a collection of local computational devices distributed through a physical space, in which distance matters and where the system should explicitly using the concept of space in computations. Accurate handling of location information in IoT is being built on the standards for location well established by several standards developing organizations, in particular as established by the Open Geospatial Consortium (OGC) [11].

In 2012 OGC members established the Sensor Web for Internet of Things Standards Working Group and started the development of the SensorThings API. Developed based on the existing OGC Sensor Web Enablement (SWE) standards, the SensorThings API is a new light-weight standard designed specifically for IoT devices and applications. The existing OGC SWE standards enable all types of sensors and actuators discoverable, accessible and re-useable via the Web. These standards have been widely implemented around the world. SWE standards, however, are as complex as necessary to support tasks such as controlling Earth imaging satellites and archiving national libraries of geological observation data, and thus are, too "heavyweight" for the resource-constrained IoT applications. The OGC SensorThings API can be considered as a lightweight SWE profile suited particularly for IoT applications. As a result, the OGC SensorThings API is a new and efficient API based on the proven and widely implemented SWE standard framework.

The OGC SensorThings API is currently a standard candidate and has been released for public review. A summary of the current SensorThings API is described as follows. The current SensorThings API candidate consists of two layers of standards for connecting various types of IoT sensing devices. Each standard layer deals with a 'level of interoperability' issue. The first layer is the *IoT Resources Model Layer* that enables the understanding and use of heterogeneous IoT devices, their sensing and control capabilities, and associated metadata. This layer consists of the standards based data model describing the entities (*i.e.*, Resources in the Resource-Oriented Architecture) and their relationships. The second layer is the *IoT Service Interface Layer* that defines (1) the URI patterns for IoT resource addressing, (2) the CRUD (CREATE, READ, UPDATE, and DELETE) operations capable of being performed against the IoT Resources, and (3) the available query parameters for filtering the IoT resources.

Figure 4.3 illustrates the SensorThings IoT Resources Model. It has two profiles, namely the sensing profile (Right figure) and the control profile (Left figure). The sensing profile consists of the resources that allow users and applications to understand the data collected by the IoT sensors. The control profile consists of the resources that allow users and applications to send tasks and control the IoT actuators.

The core of the SensorThings resource model is a *Thing*. SensorThings API uses ITU's definition [27], *i.e.*, a *Thing* is an object of the physical world (physical things) or the information world (virtual things) that is capable of being identified and integrated into communication networks. Every *Thing*

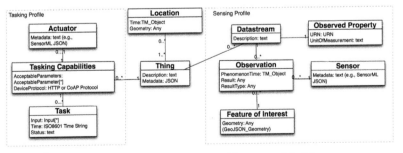

Figure 4.3   SensorThings IoT Resources Model

has zero to many **Locations**. Each **Thing** can have zero to many **Datastreams**, and **Datastream** forms the core of the sensing profile. The sensing profile is based on the standard O&M data model [28], *i.e.*, an **Observation** is modeled as an event performed by a **Sensor** (or **Process**) at a **Location** and a **Time** that produces a result whose **Value** is an estimate of an **Observed Property** of the **Feature of Interest**.

The left hand-side of Figure 4.3 illustrates the SensorThings API's control profile. A controllable **Thing** can have zero to many **Tasking Capabilities** that accept certain **AccetableParameters** allowing users to compose and send feasible **Tasks** that can be performed by an **Actuator**. The control profile is based on the OGC Sensor Planning Service standard [29]. The main difference is that SPS uses a Service-Oriented Architecture and the SensorThings API uses a Resource-Oriented Architecture.

The SensorThings IoT service interface consists of the following three components: (1) the URI patterns for IoT resources addressing, (2) the CRUD operations capable of being performed on the IoT resources, and (3) the available query parameters for filtering the IoT resources.

In order to perform CRUD action on the Resources, the first step is to address the target resource(s) through their URI. Figure 4.4 shows the three URI components defined by RESTful IoT, namely the service root URI, the resource path, and the query options. The service root URI is the location of the SensorThings service. By attaching the resource path after the service root URI, users can address to the Resources available in a SensorThings service. And when users perform a READ action on Resources, users can apply query options to further process the addressed resources, such as sorting by properties and filtering with criteria.

A SensorThings service will group the same types of entities into collections. Each entity has a unique identifier and one to many properties. In the case of an entity holding a relationship with entities in another collection, this entity has a navigation property (*i.e.*, a link) linking to other entities. The navigation property enables users to access the resources with a multi-facet-based structure rather than a hierarchical structure. This multi- facet-based design is based on the OASIS OData standard specification [30].

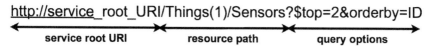

Figure 4.4   URI components defined by RESTful IoT

```
GET http://../                              GET http://../Things(1)

{                                           {
"Collections": [                                Description: "This is an air quality pig.",
{"uri": "http://.../Things"},                   Self-Link: "http://-/Things(1)",
{"uri": "http://.../Locations"},                Locations:
{"uri": "http://.../Datastreams"},              {
{"uri": "http://.../Observations"},                 Association-Link: "Things(1)/$links/Locations",
{"uri": "http://.../FeaturesOfInterests"},          Navigation-Link: "Things(1)/Locations"
{"uri": "http://.../ObservedProperties"},       },
{"uri": "http://.../TaskingCapabilities"},      ID: 1,
{"uri": "http://.../Tasks"}                     Datastreams:
]                                               {
}                                                   Association-Link: "Things(1)/$links/Datastreams",
                                                    Navigation-Link: "Things(1)/Datastreams"
                                                },
                                                TaskingCapabilities:
                                                {
                                                    Association-Link: "Things(1)/$links/TaskingCapabilities",
                                                    Navigation-Link: "Things(1)/TaskingCapabilities"
                                                }
                                            }
```

Figure 4.5   GET samples

Once a resource can be identified by an URI, CRUD actions (HTTP methods of POST, GET, PUT, and DELETE) can be performed on the resource. Figure 4.5 shows two GET examples. The left hand side shows an example of a SensorThings service root, *i.e.*, all collections of a SensorThings service instance. The right hand side shows the instance of a *Thing*, and it can be retrieved by issuing a GET request to the URI path of the *Thing*.

The latest SensorThings API draft is available at http://ogcnetwork.net /sensorthings. And at the moment the SWE-IoT SWG is seeking public comments and will consider all comments when preparing a final draft of the candidate standard. The SW-IoT SWG will consider all comments when preparing a final draft of the candidate standard. The SW-IoT SWG plans to submit the final draft to the OGC Technical Committee for approval in 2014.

4.3.8 oneM2M

The oneM2M Partnership Project "oneM2M" [31] brings together the leading Information and Communications Technologies (ICT) Standards Development Organisations from around the world. The seven founding oneM2M partners Type1 working together with ETSI - European Telecommunications Standards Institute, are: ARIB - Association of Radio Industries and Businesses (Japan), ATIS - Alliance for Telecommunications Industry Solutions (US), CCSA - China Communications Standards Association (China), TIA - Telecommunications Industry Association (US), TTA - Telecommunications Technology Association (Korea), and TTC - Telecommunication Technology Committee of Japan (Japan).

In addition, oneM2M has welcomed other industry organizations as partners, including as partners Type2: the Broadband Forum (BBF), the Continua

Health Alliance, the HGI (Home Gateway Initiative), and the Open Mobile Alliance (OMA). At the close of its first year and a half, oneM2M has over 260 member companies from around the world, and has conducted plenary meetings in Europe, China, the U.S., Korea, Canada, and Japan.

Launched in July 2012, oneM2M is committed to unifying the global M2M community by developing a cost-effective, widely available service layer that meets the needs of both the communications industry and vertical industry members. oneM2M welcomes the opportunity to collaborate with other industry organizations as well as vertical market segments in the M2M space to extend interoperability, and enhance security and reliability by reducing industry fragmentation.

oneM2M is governed by a Steering Committee (SC) made up of all Partners, and is supported by Finance, Legal and MARCOM sub-committees, as well as a Methods and Procedures group. Technical work is progressed by a Technical Plenary, organized into five working groups: Requirements (WG1), Architecture (WG2), Protocols (WG3), Security (WG4), and Management, Abstraction, & Semantics (WG5).

Over the last year, within the Technical Plenary and Working groups, hundreds of technical contributions from member companies have been discussed, modified and agreed. The result is that the foundation of an initial set of oneM2M service layer requirements is nearly complete, a oneM2M architectural vision is underway, and work has begun on the path towards oneM2M protocol determination. Security and Management topics are being progressed in parallel and coordinated with all other working groups. The first technical reports issued by oneM2M were approved by the Technical Plenary in August 2013.

Looking toward the future, oneM2M is anticipating an initial release of oneM2M technical specifications in mid-2014. These documents can then be adopted and published by the founding partners for use in both global and regional M2M implementations. Subsequent oneM2M work will enhance the initial release with additional functionality and interoperability, and will result in future releases.

4.3.9 GS1

GS1 is an open, neutral, not-for-profit industry-driven standard organisation responsible for defining unique identifiers for items, parties, documents, locations, events and other "things" for more than 40 years. The GS1 standards for identification, semantics and communication are used directly by over

1.5 million companies and indirectly by billions of consumers every day. Barcodes, RFID tags and the underlying, globally-unique numbering system combined with data sharing standards offer the opportunity to dramatically enhance the efficiency of supply and demand chains.

4.3.9.1 The Role of Standards

The GS1 System is primarily concerned with raising the efficiency of business processes and providing cost savings through automation based on globally unique identification and digital information. The role of GS1 Standards is to further the following objectives [32]:

- *To facilitate interoperability in open supply chains*
  GS1 Standards include data standards and information exchange standards that form the basis of cross-enterprise exchange as well as standards for physical data carriers, i.e. bar codes and RFID tags.
- *To foster the existence of a competitive marketplace for system components*
  GS1 Standards define interfaces between system components that facilitate interoperability between components produced by different vendors or by different organisations' in-house development teams. This in turn provides choice to end users, both in implementing systems that will exchange information between trading partners and in those that are used entirely internally.
- *To encourage innovation*
  GS1 Standards define interfaces, not implementations. Implementers are encouraged to innovate in the products and systems they create, while interface standards ensure interoperability between competing systems.

4.3.9.2 GS1 Standards: Identify, Capture, Share

GS1 Standards may be divided into the following groups according to their role in supporting information needs related to real-world entities in supply chain business processes [32]:

- Standards which provide the means to **Identify** real-world entities so that they may be the subject of electronic information that is stored and/or communicated by end users. Real-world entities include trade items, logistics units, legal entities, physical locations, documents, service relationships, etc.
- Standards which provide the means to automatically **Capture** data that is carried directly on physical objects, bridging the world of physical

things and the world of electronic information. GS1 data capture standards currently include definitions of bar code and radio-frequency identification (RFID) data carriers which allow GS1 Identification Keys and supplementary data to be affixed directly to a physical object, and standards that specify consistent interfaces to readers, printers, and other hardware and software components that connect the data carriers to business applications.

- Standards which provide the means to **Share** information, both between trading partners and internally, providing the foundation for electronic business transactions, electronic visibility of the physical and digital world, and other information applications. GS1 standards for information sharing include data standards for master data, business transaction data, and physical event data, as well as communication standards for sharing this data between applications and trading partners.

Figure 4.6 gives a high-level overview of GS1 standards.

4.3.9.3 Looking forward

GS1 has seen massive adoption of unique instance identification and EPC-enabled Radio-Frequency Identification (RFID) technologies driven by a need for inventory management accuracy and fight against theft. The recently

Figure 4.6   GS1 standards

released Gen2v2 specification for EPC-enabled RFID has set the standard for expansions of RFID tag capability from traditional locate/read applications to fully-interactive locate/read/access/write/authenticate applications. Such applications will have far-reaching implications to consumer privacy, anti-counterfeiting, security, and loss prevention.

In the fields of pharmaceuticals and medical devices, we are seeing a significant increase of item identification at the instance level (represented in both barcodes and RFID) and in plans to share information about custody of items along the supply chains using the Internet and GS1 standard applications (Electronic Product Code Information Services - EPCIS).

Such combinations of GS1 technologies are foundational examples of the power of the Internet of Things: consistent identification of things for representation on open networks, consistent communication about (and by) things, and robust discovery services for information that has been shared about things.

In the future, there will be a significant increase in web-based applications developed by industry that are focused on improving the consumer experience. Standards will be required to better enable these new applications. A critical issue is further defining the data standards for various APIs built to provide better service for consumers. Common vocabularies are critical, but how to most clearly define the data that needs to be standardized for these applications in various domains of use is of paramount importance.

4.4 IERC Research Projects Positions

4.4.1 BETaaS Advisory Board Experts Position

IoT is shaping the evolution to a ubiquitous Internet connecting people and heterogeneous things, seamlessly integrated, anytime and everywhere. This require scalability, resilience, security, interworking between systems of systems, autonomous and trusted self-organizing networks of systems, ad hoc power consumption, and 'intelligence' (smart services). From applications down to real world there are two 'semantic interoperability' challenges, we hope we could start with a consensus on following two requirements (pre-standardization) [33]:

1. at the highest level: we lack of common semantic IoT domain of definition with a common structured and a common method to describe things (real, virtual, human, aggregated), associated things and services, events and types of operations at highest semantic level

2. at the lowest level: we lack of shared pre-build real examples of semantic things objects events operations to make the adoption of semantic things interoperability more easy to understand and to implement

If different implementers of IoT as a Service could agree on such common requirements and to evolve their own solutions to an open semantically interoperable IoTaaS then the IoT of the science fiction movies can become reality. IoTaaS social/market adoption and fair approach between technology push and market demand requires a pre-standardization to build consensus on the vision and requirements and to evolve from today's IoT/M2M legacy. BETaaS is making an effort to approach standardization, proposing a solution that tries to overcome at least the first of these two limitations. This solution is based on WordNet, a lexical database in English that groups English words into sets of synonyms, and which defines semantic relations between these sets of synonyms. Given a thing description from the IoT, BETaaS uses WordNet to infer information about its location and its type. In the same way, given an application requirement, BETaaS uses WordNet to infer information about the type of things demanded and their location.

4.4.2 IoT6 Position

The IoT6 European research project [34] is researching the potential of IPv6 and related standards for the Internet of Things. It has disseminated its results with and contributed to several international standardizations bodies, including in the IETF, IEEE, ITU-T and OASIS. The projects' results are confirming the importance and relevance of IPv6 to enable a global Internet of Things. IPv6 is not only providing a large scale addressing scheme and a native integration with the worldwide Internet, but also a source of many relevant and useful features, including self-configuration mechanisms and secured end-to-end connections. IoT6 clearly supports an extended use of IPv6 for the Internet of Things interconnections.

The public IPv4 address space managed by IANA [12] has been completely depleted by Feb 1st, 2011. This creates by itself an interesting challenge when adding new things and enabling new services on the Internet. Without public IP addresses the Internet of Things capabilities would be greatly reduced. Most discussions about IoT have been based on the illusionary assumption that the IP address space is an unlimited resource or it is even taken for granted that IP is like oxygen produced for free by nature. Hopefully, the next generation of Internet Protocol, also known as IPv6 brings a solution. In early 90s, IPv6 was designed by the IETF IPng (Next Generation) Working Group and promoted by the same experts within the IPv6 Forum since 1999. Expanding the IPv4 protocol suite with larger address space and defining new capabilities restoring end to end connectivity, and end to end services, several IETF working groups have worked on many deployment scenarios with transition models to interact with IPv4 infrastructure and services. They have also enhanced a combination of features that were not tightly designed or scalable in IPv4, such as IP mobility and ad-hoc services, catering for the extreme scenario where IP becomes a commodity service enabling lowest cost networking deployment of large scale sensor networks, RFID, IP in the car, to any imaginable scenario where networking adds value to commodity. For that reason, IPv6 makes feasible the new conception of extending Internet to consumer devices, physical systems and any imaginable thing that can be benefited of the connectivity. IPv6 spreads the addressing space in order to support all the emerging Internet-enabled devices. In addition, IPv6 has been designed to provide secure communications to users and mobility for all devices attached to the user; thereby users can always be connected. This work provides an overview of our experiences addressing the challenges in terms of connectivity, reliability, security and mobility of the Internet of Things through IPv6.

The Position Paper "Internet of Everything through IPv6" [35] has been used as a reference for this section. This paper describes the key challenges, how they have been solved with IPv6, and finally, presents the future works and vision that describe the roadmap of the Internet of Things in order to reach an interoperable, trustable, mobile, distributed, valuable, and powerful enabler for emerging applications such as Smarter Cities, Human Dynamics, Cyber-Physical Systems, Smart Grid, Green Networks, Intelligent Transport Systems, and ubiquitous healthcare.

4.5 Conclusions

Most Internet standards are too complex for the constrained devices in the IoT and many of these devices are designed to run proprietary protocols, creating data silos. In the short run the vertical integration of sensors and business services will dominate IoT. As wireless sensors are deployed, each of them using different standards/protocols, services providers arise to collect and interpret disparate data, and standards are need to ensure that is possible.

More and more hardware companies push for standardization so they can capitalize on services revenue since many of them see beyond the "things" and focus on the services built on the "Internet of Things".

There is a good momentum on IoT standardisation and IERC and its participating projects are seen as a catalyst and an European IoT coordination platform facilitating international world-wide dialog. IoT Workshops co-organised between the European Commission, IoT Research and Innovation projects, IoT Industry Stakeholders and IoT Standard Organisation groups are continuing.

These workshops facilitate interoperability testing events to stimulate IoT community building to reach consensus on IoT standards common developments on all protocol layers.

New domains have to be integrated into the overall view like the standardisation development in ITS (Intelligent Transport Systems) in ETSI and ISO. A significant effort will be required to come to an overall cross vertical IoT vision and interoperable standards environments. In this section an overview over the European and world-wide IoT standardization landscape has been given. It represents only a part of the activities in the domain and is by no mean a comprehensive full coverage of all IoT related standards activities. Several additional groups are active in the domain or started to enter the IoT working field.

But already this overview depicts the vast number of different organizations and applications related to the future IoT. It also demonstrates the significant need of strong coordination between these activities in order to push for a horizontally integrated IoT ecosystem. IoT is not a single system and not only one standard will define IoT in the future. Interoperability between the domains and systems will be a key factor for the sustainable success of IoT.

References

[1] ITU-T, Internet of Things Global Standards Initiative, http://www.itu.int/en/ITU-T/gsi/iot/Pages/default.aspx
[2] International Telecommunication Union — ITU-T Y.2060 — (06/2012) — Next Generation Networks — Frameworks and functional architecture models — Overview of the Internet of things.
[3] Vermesan, O., Friess, P., Guillemin, P., Gusmeroli, S., et al., "Internet of Things Strategic Research Agenda", Chapter 2 in O. Vermesan and P. Friess (Eds.), *Internet of Things—Global Technological and Societal Trends*, River Publishers, Aalborg, Denmark, 2011, ISBN 978-87-92329-67-7.

[4] Guillemin, P., Berens, F., Carugi, M., Arndt, M., et al., "Internet of Things Standardisation - Status, Requirements, Initiatives and Organisations", Chapter 7 in O. Vermesan and P. Friess (Eds.), *Internet of Things: Converging Technologies for Smart Environments and Integrated Ecosystems*, River Publishers, Aalborg, Denmark, 2013, ISBN 978-87-92982-73-5.

[5] eCall - eSafety Support, online at http://www.esafetysupport.org/en/ecall_toolbox/european_commission/index.html

[6] European Commission, "Smart Grid Mandate, Standardization Mandate to European Standardisation Organisations (ESOs) to support European Smart Grid deployments", M/490, Brussels 1st March, 2011.

[7] Global Certification Forum, online at http://www.globalcertification forum.org

[8] Colitti, W., Steenhaut, K., and De Caro, N., "Integrating Wireless Sensor Networks with the Web," Extending the Internet to Low Power and Lossy Networks (IP+SN 2011), 2011 online at http://hinrg.cs.jhu.edu/joomla/images/stories/IPSN_2011_koliti.pdf.

[9] FG M2M, http://www.itu.int/en/ITUT/focusgroups/m2m/Pages/default.aspx

[10] ITU-T workshop "IoT – Trends and Challenges in Standardization", http://www.itu.int/en/ITU-T/Workshops-and-Seminars/iot/201402/Pages/Programme.aspx

[11] The Open Geospatial Consortium: http://www.opengeospatial.org/

[12] IANA, http://www.iana.org

[13] Final SmartHouse Roadmap Recommendations to the European Commission, online at ftp://ftp.cencenelec.eu/CENELEC/SmartHouse/SmartHouseRoadmap.pdf

[14] SmartHouse Roadmap, leaflet, online at ftp://ftp.cencenelec.eu/CENELEC/SmartHouse/SmartHouseBrochure.pdf

[15] ETSI, "work programme 2013-2014", online at http://www.etsi.org/images/files/WorkProgramme/etsi-work-programme-2013-2014.pdf

[16] ETSI, "ETSI Long Term Strategy", online at http://etsi.org/WebSite/document/aboutETSI/LTS%20Brochure%20W.pdf

[17] ETSI, "annual report 2012", published April 2013, online at http://www.etsi.org/images/files/AnnualReports/etsi-annual-report-april-2013.pdf

[18] "The Standard", ETSI Newsletter, September 2013, online at http://www.etsi.org/Images/files/ETSInewsletter/etsinewsletter_sept 2013.pdf

[19] Terms of Reference (ToR) for Technical Committee , "Smart M2M", online at https://portal.etsi.org/SmartM2M/SmartM2M_ToR.asp

[20] IEEE Standards Association, "Internet of Things", online at http://standards.ieee.org/innovate/iot/index.html

[21] IEEE Standards Association, "IEEE Standards Activities in Cloud Computing", updated 20 June 2013, online at http://standards.ieee.org/develop/msp/cloudcomputing.pdf

[22] IEEE Standards Association, "IEEE Standards Activities in the eHealth Space", updated 19 June 2013, online at http://standards.ieee.org/develop/msp/ehealth.pdf

[23] IEEE Standards Association, "IEEE Standards Activities in the eLearning Space", updated 20 June 2013, online at http://standards.ieee.org/develop/msp/elearning.pdf

[24] IEEE Standards Association, "IEEE Standards Activities in th Intelligent Transportation Systems (ITS) Space (ICT Focus), updated 27 June 2013, online at http://standards.ieee.org/develop/msp/its.pdf

[25] IEEE Standards Association, " IEEE Standards Activities in the Network and Information Security (NIS) Space", updated 19 June 2013, online at http://standards.ieee.org/develop/msp/nis.pdf

[26] IEEE Standards Association, "IEEE Standards Activities in the Smart Grid Space (ICT Focus)", updated May 2013, online at http://standards.ieee.org/develop/msp/smartgrid.pdf

[27] ITU, Y.2060: Overview of the Internet of things, http://www.itu.int/rec/T-REC-Y.2060-201206-P

[28] OGC, "Observations and Measurments", online at http://www.opengeospatial.org/standards/om

[29] OGC, "Sensor Planning Service (SPS)", online at http://www.opengeospatial.org/standards/sps

[30] OASIS, "OASIS Open Data Protocol (OData) TC", online at https://www.oasis-open.org/committees/tc_home.php?wg_abbrev=odata

[31] oneM2M, http://www.onem2m.org/

[32] The GS1 System Architecture, Issue 3.0, 14 April 2014, http://www.gs1.org/docs/gsmp/architecture/GS1_System_Architecture.pdf

[33] EU-China FIRE Advisory Board, Patrick Guillemin, NFV, SDN, AFI, "Future Internet Standardisation", 18.04.2014, Beijing. online at http://www.euchina-fire.eu/wp-content/uploads/2014/04/ECIAO_conf1_Patrick-Guillemin-Introduction-of-Future-Interent-Standard.pdf

[34] IoT6 European research project, www.iot6.eu

[35] Jara, A.J., Ladid, L., and Skarmeta, A., "Internet of Everything through IPv6: An Analysis of Challenges, Solutions and Opportunities", online at http://www.ipv6forum.com/iot/index.php/publications/9-uncategorised/84-ioe-positionpaper

[36] CEN BOSS, "Technical Structures", online at http://boss.cen.eu/TechnicalStructures/Pages/default.aspx

[37] CEN, "Technical Bodies", online at http://standards.cen.eu/dyn/www/f?p=CENWEB:6:::NO:::

[38] CENELEC, "European Standards (EN)", online at http://www.cenelec.eu/standardsdevelopment/ourproducts/europeanstandards.html

[39] CENELEC, "Technical Specifications", online at http://www.cenelec.eu/standardsdevelopment/ourproducts/technicalspecification.html

[40] CENELEC, "Technical Reports (TR)", online at http://www.cenelec.eu/standardsdevelopment/ourproducts/technicalreports.html

[41] CENELEC, "CENELEC Workshop Agreements (CWA)", online at http://www.cenelec.eu/standardsdevelopment/ourproducts/workshopagreements.html

[42] CEN, "CEN Members", online at http://standards.cen.eu/dyn/www/f?p=CENWEB:5

[43] CEN, "CEN/TC 225 – AIDC technologies, General", online at http://standards.cen.eu/dyn/www/f?p=204:7:0::::FSP_ORG_ID:6206&cs=1E12277AECC001196A7556B8DBCDF0A1C

[44] CEN, "CEN/TC 225 – AIDC technologies, Structure", online at http://standards.cen.eu/dyn/www/f?p=204:29:0::::FSP_ORG_ID,FSP_LANG_ID:6206,25&cs=136D1799132ED1E13E56D38C2E645A7D2#1

[45] CEN, "CEN/TC 225 – AIDC technologies, Work programme", online at http://standards.cen.eu/dyn/www/f?p=204:22:0::::FSP_ORG_ID,FSP_LANG_ID:6206,25&cs=136D1799132ED1E13E56D38C2E645A7D2

[46] CENELEC, "CENELEC Products", online at http://www.cenelec.eu/standardsdevelopment/ourproducts/

[47] CENELEC, "Smart grids", online at http://www.cenelec.eu/aboutcenelec/whatwedo/technologysectors/smartgrids.html

[48] CENELEC, "Smart grids", online at http://www.cencenelec.eu/standards/Sectors/SustainableEnergy/Management/SmartGrids/Pages/default.aspx

[49] CENELEC, "CENELEC project SmartHouse Roadmap", online at http://www.cenelec.eu/aboutcenelec/whatwedo/technologysectors/smarthouse.html

[50] CENELEC, "Electric Vehicles", online at http://www.cenelec.eu/about
cenelec/whatwedo/technologysectors/electricvehicles.html

[51] Wikipedia,"ISO/IEC JTC1/SWG5", online at http://en.wikipedia.org
/wiki/ISO/IEC_JTC_1/SWG_5

[52] CENELEC,online at http://www.cenelec.eu/

[53] ETSI, online at http://www.etsi.org/

[54] CEN/CENELEC Internal Regulations – Part 2: Common Rules for
Standardization Work, 2013, http://boss.cen.eu/ref/IR2_E.pdf

[55] CEN-CENELEC GUIDE 20, Edt. 3, 2013-07, ftp://ftp.cencenelec.eu/EN
/EuropeanStandardization/Guides/20_CENCLCGuide20.pdf

[56] CENELEC, Smart grids, online at http://www.cencenelec.eu/standards/
Sectors/SustainableEnergy/Management/SmartGrids/Pages/default.aspx

[57] Wikipedia, "ISO/IEC JTC1/WG7", online at http://en.wikipedia.org
/wiki/ISO/IEC_JTC_1/WG_7#cite_note-11

[58] ETSI, http://www.etsi.org/images/articles/etsiclusters2.png

[59] Project SUNRISE, online at www.sunrise-project.eu

[60] Open GNSS Service Interface Forum, online at www.opengnssforum.eu

[61] ITU-T JCA-IoT (Joint Coordination Activity on Internet of Things,
online at http://www.itu.int/en/ITU-T/jca/iot

[62] IoT Standards Roadmap, online at http://itu.int/en/ITU-T/jca/iot/
Documents/deliverables/Free-download-IoT-roadmap.doc

5

Dynamic Context-Aware Scalable and Trust-based IoT Security, Privacy Framework

Ricardo Neisse,[1] Gary Steri,[1] Gianmarco Baldini,[1] Elias Tragos,[2]
Igor Nai Fovino[1] and Maarten Botterman[3]

[1]*Joint Research Centre, European Commission, Italy*
[2]*Foundation for Research and Technology Hellas (FORTH), Greece*
[3]*GNKS, The Netherlands*

5.1 Introduction

The evolution of Internet toward Internet of Things (IoT) will have a major impact on the lives of citizens as new services and applications can be developed by the integration of the physical and digital worlds. Mobiles, wearable sensors, and "smart" devices with improved capabilities to act autonomously can be used to support new applications for healthcare, transportation and energy savings, improve business efficiency, enhance security or, in general, to support the needs of the citizen.

The Internet of Things was said to be first quoted by Kevin Ashton in 1999 [1]. "Things" are known to have been connected pretty early, such as the camera observing the coffee pot in the Trojan Room within the computer laboratory of the University of Cambridge, installed in 1991, or the Coke Machine polling at Carnegie Mellon's Computer Science department in 1982. But it is only in recent years that the interest for the Internet of Things has risen to high level, and predictions go up to an expectation of 50 billion devices that will be connected by 2020 [2]. In recent years, various definitions of IoT have been presented by various sources. The International Telecommunication Union (ITU) Internet report [3] focused on the connectivity aspects of IoT in various domains: "from anytime, anyplace connectivity for anyone, we will now have connectivity for anything". In a similar way, ETSI [4] has defined IoT as "The Internet of Things allows people and things to be connected Anytime, Anyplace, with Anything and Anyone, ideally using Any path/network and Any service".

Beyond connectivity, one of the other features of IoT is the capability of embedding intelligent behaviour in the "things", which can be sensors or actuators. Thus, the adjective "smart" is often seen in IoT references: smart home, smart city, smart car and so on. The concept is to use the increased connectivity provided by wireless communication technologies, the increased computing power and memory capacity of embedded devices to implement autonomous behaviour, which can support and augment the citizen capabilities. It enables new services, and new ways to offer services that already exist.

On the other side, the increasing amount of data originated by the IoT objects can pose serious threats to the privacy and security of the citizen, because, for example, the activities of a citizen can be tracked at any time and place. While there can be contexts where this may not be an issue and it is actually a benefit or it is specifically requested (e.g. citizen at work or healthcare support to an elderly person, or in emergency situations), it should respect the fact that the citizen has the right to his or her own privacy.

In other contexts, the security of the operations performed by the citizen is also necessary. For example, the increasing adoption of wireless technology for payments or the activation of various services through authentication require the design and implementation of security solutions. The solutions designed to support security and privacy needs should be able to support different contexts (e.g., at home or at work) and to be scalable/interoperable for the increasing number of IoT devices which the citizen interacts with.

It is also clear that the perception of "privacy" and the trade-off between (personal) privacy and (societal) security is not a fixed concept, but a moving target that is the result of experiences in society, and the ability of citizens to understand what is going on, and to make choices: a clear policy issue.

In addition, the "value" of data, now emerging at the heart of new business models, will further develop the "hunger" for data, and the Internet of Things will be a main contributor to the amount of generated data.

Getting in place a clear framework that facilitates the "responsible use" of data from a privacy and data protection perspective is of the highest importance.

Various challenges have been identified [5, 6] to support security and privacy in the evolution of IoT:

- markets won't invest in right level of security as today "time to market" is a bigger driver than the level of security or privacy, today;
- the definition of privacy by regulatory bodies can be quite different among different geo-political zones;
- security solutions are usually designed to protect business data in vertical applications. As a consequence, they may be difficult to be extended to other applications or devices;
- the deployment of numerous devices with limited processing and memory capabilities can increase the threat space of the IoT applications. In other words, an attacker can exploit a weakness in an IoT device with limited capability to penetrate connected IoT applications, which are supposedly considered more secure. In addition, "things" that can act (e.g., actuators) on the physical or digital world can become new end points for attack – either by tampering with the "thing" directly, or by providing the capabilities for more sophisticated threats;
- the need to protect data in IoT is in opposition to the market drivers to generate and access the vast amount of data generated by IoT devices for commercial applications such as targeted advertising and Location Based Services;
- the requirement for enabling the reuse of IoT data gathered for one application towards other applications is mainly contradicting with privacy and especially privacy-by-design.

It is important to identify these issues and challenges, but it is even more important to research and define solutions at this current phase, where IoT technologies are in the way to be defined and deployed. The research community has been investigating the security and privacy aspects of IoT with growing interest and a survey of the current research activities and the related results is presented in this book chapter.

Additionally, we will also focus on a definition of a framework to support security and privacy in IoT, which is based on the results from the FP7 projects involved in the Internet of Things Cluster (IERC) [7]. The main element of the framework is a usage control toolkit, where policies can be used to define the access to data and resources in IoT, with the possibility of supporting dynamic changes of context. In other words, the policies can be defined for different contexts (e.g., work, personal life), for different roles and different types of

IoT devices. The toolkit is complemented by other elements to address the challenges described above. The framework is applied to a smart city scenario focusing on the interaction between a smart home, a smart vehicle, and a smart office in order to demonstrate the feasibility and the deployment challenges.

5.2 Background Work

An extensive survey on frameworks for Context Aware Computing for the Internet of Things is presented in [8]. The survey defines the main context features, which are desirable in the framework, and identifies a large number of frameworks from research and commercial projects, which supports these features to some degree. One of the first examples is the Context Toolkit described in [9], which has the objective to facilitate the development of context-aware applications. The design of the Context Toolkit is based on three main elements: (i) the context widget with interfaces to the sensors, (ii) the context interpreter to process and analyze the data from the sensors and (iii) the context aggregator, which aggregates the data to support the application. While the Context Toolkit has presented some of the initial concepts to support Context-based applications, the security and privacy aspects were not fully addressed.

The Context Broker Architecture (CoBrA) [10] is one of the frameworks which addresses security and privacy aspects. In [10], the framework is applied to a smart meeting room system, where the confidentiality of the data distributed in the "room" and the privacy of the users participating to the meeting is of primary importance. The paper acknowledges the difficulty to protect privacy when the context can be dynamic and the users must manually define the privacy policies for each context. In addition, users may not be aware that data provided by them are used in some other context or domain by the application. The paper suggests the adoption of the Standard Ontology for Ubiquitous and Pervasive Applications (SOUPA) to define the access to data on the basis of semantic information. While the approach has merits, the authors recognize that this approach could not be flexible enough to support a Dynamic Context and it may not address privacy concerns such as the logging and persistent storage of a user' s private information by the agents.

More recently, Gessner et al. [11] have proposed a set of trust-enhancing security functional components based on Identity Management (IM), Authorization (AuthZ), Key Exchange and Management (KEM) and trust and reputation management (TRA). These components are linked to provide a framework for security and privacy in IoT. AuthZ is based on an Access Control

Model, where policies can be defined. This is a similar but simpler approach to the framework presented in this book chapter. Pseudonyms are also used to protect the privacy of the users. While the definition of the components is sound, the paper does not address dynamic change of contexts, which is an important element in IoT.

The authors in [12] present a framework to empower the users to control the generation and access to their personal data. The framework is based on three main components: (i) User Controlled Privacy Preserved Access Control Protocol to regulate the transmission of personal data, (ii) a k-anonymity solution to anonymize the data of the users, which can be regulated on the basis of the users profile or the context and (iii) additional privacy solutions for stored data based on default privacy protection levels. The combination of these components can support the privacy of users from the generation of users'data to the storage of data on the basis of the profile of the users or the context. This is an alternative approach to what proposed in this book chapter but with similar objectives.

5.3 Main Concepts and Motivation of the Framework

As already claimed in the introduction, the massive adoption of the IoT paradigm in the daily life poses serious questions under a privacy and security perspective. IoT devices are today disseminated everywhere; in smart-houses, sensors connected to the Internet are used to monitor the environment (e.g. IP-cameras, temperature sensors, motion sensors, smart-meters, etc.), and on the basis of the information collected, the status of the environment is modified through actuators.

In the same way, in smart-cities, IoT devices are used to monitor the city-state evolution and to eventually operate to vary the state. IoT applications for smart cities span from environmental monitoring, traffic monitoring and management, smart parking, smart lighting, waste management, surveillance, safety and emergency alarms. Sensors monitoring the traffic evolution in the streets can trigger modification in the semaphores' temporization to solve traffic jams. Traffic lights can take smart decisions and cooperate with each other to change the green/red light durations according to the traffic on the roads. Lights at the streets can be adaptive towards minimizing energy consumption. Smart waste-bins may inform the public servants when they should be emptied. These are only few examples of the many benefits of IoT in city-wide areas.

The general implication of this picture is that today, our environment is disseminated by objects which are potentially tightly linked to our life and which, if not strictly regulated, can easily infringe the security and privacy of the citizen. For example, traffic cameras monitoring a street may capture videos and images of people passing by on the pavement. This information, if not properly protected, may become available to third parties, and thus private information on the location of citizens at some point in time can be disclosed to unauthorized persons. Similarly, location information sent by mobile phones of users (while in their cars) assisting on traffic monitoring (via a crowdsourcing application), even if sent anonymously, can be easily mapped to a specific person and can reveal user movements and habits. Additionally, in crowdsourcing applications, malicious users may easily transmit false information affecting the decisions of the system. For example, in the previous scenario of "smart" traffic lights, a user may send false information regarding traffic so that the traffic light becomes green and he gets faster to his destination. There are many other similar scenarios that justify the importance of security and privacy in IoT based applications. Security and Privacy are themselves the two macro challenges in IoT environments, and they can be split and detailed in several particular challenges listed here in the following paragraphs.

5.3.1 Identity Management

According to the standard ISO/IEC 24760 [13] a digital identity is defined as a set of attributes related to an "entity", which refers to an individual, an organization, or a device. Attributes are properties of the entity (e.g. address, phone number etc.). The digital identity definition has been extended recently with a sort of "inheritance principle" regarding the IoT world. To get access to more and more complex online services IoT devices need to be configured by their owners using their own credentials, giving to these devices rights to operate in their name. Let us take as example a smart-TV: a citizen that wants to download and see online content should provide the smart-TV with a mean to authenticate itself to the online services. Typically, the authentication will imply the use of some sort of digital-identity linked to the owner of the TV-subscription; in other words, the smart-TV inherits a "portion" of the identity of its owner. The same situation happens when, for example, the citizen configures his mobile-phone to get synchronized with the company's calendar. To get direct access to this commodity, the smart-phone will need to authenticate itself to the calendar service using some personal credentials;

again, the smart-device inherits part of the identity of its owner. The same principle can be applied considering the more extended scenario of a Smart City, where digital identities or aggregates of digital identities are associated to complex systems used to deliver secure and trusted physical services to the citizen, e.g. public transportation, car to car communication, remotely monitored Health care devices etc. However, digital identities do not impact only on the daily life of the citizen, as their role is becoming more and more important also in the industrial sector. Let us consider the world of Industrial Control Systems (ICS); the increasing use of general purpose telecommunication networks (i.e. Internet) in these infrastructures, acted as a sort of glue, so that, today, we can say that ICS (and SCADA systems) are remotely controlled and accessed. Also in this case digital identities have a relevant role. To access certain remote components or control servers, identities with associated roles and rights need to be used. Their management, the way in which they are protected and revoked – if needed, should and must be one of the top priorities for the security of a critical infrastructure. The same consideration can be done also when thinking about the communication of low level control devices (e.g. PLCs). In this case, especially for those installations spread in geographically remote locations, with scarce or non-existing surveillance (for example a gas or oil pipeline passing through remote regions of the world), the problem of securely managing their digital identities (in this case crypto-material allowing to sign and authenticate their readings and control messages) should be of high relevance. An interesting playground where citizen identities and industrial infrastructures are quickly converging is that of smart-metering. Smart-meters can be considered the ultimate leafs of the smart-grids. These objects are at the moment those in charge for measuring the energy consumptions of the citizen, and, in some countries, for measuring also the energy production of the citizen. However, to really benefit from the establishment of a smart-energy grid, soon these meters will need to get more and more integrated, on a side, with the energy-distribution infrastructure, and on the other, with the citizen's home digital infrastructure. Here again the digital identity inheritance principle described above will play a relevant role in the protection of the privacy of the citizen while guaranteeing the provisioning, in a secure way, of services allowing to improve the optimization of the energy consumption and production. The challenge here is to provide a framework able to manage the identities of the different objects, while at the same time guaranteeing the right amount of information disclosure, privacy and service access permissions.

5.3.2 Size and Heterogeneity of the System

The IoT world is, by definition, an "integrated system", where different "things" interact by exchanging information and commands. These objects might be heterogeneous in terms of minimum level of security and privacy guaranteed, technology, protocol of communication and policy enforcement. Here the challenge is more related to the need for a horizontal framework able to manage security and privacy specifications in a unique and homogeneous way. These specifications will need indeed to be instantiated on "entities" potentially having completely different implementation, specifications and communication interfaces.

5.3.3 Anonymization of User Data and Metadata

Data gathered by IoT devices can be, potentially, extremely sensitive. Hence, the definition of methods and approaches allowing to identify, on the basis of a given context, what the IoT device can release in term of information became paramount. Data anonymization has been used to hide the identity of the user in the data he sends (e.g. in crowdsourcing applications) by transforming the sensitive data into data that cannot be readable by humans, and thus can be easily sent within a system/network without having the risk of being disclosed to unauthorized third parties. In a similar way, the pseudonymization of user data is also used in various systems, for replacing the most identifying fields of user data with one or more pseudonyms (artificial identifiers). These methods are considered to be the first step towards retaining a minimum level of user privacy.

5.3.4 Action's Control

IoT devices might take actions (e.g. trigger actuators) on the base of a context. These actions might regard not only physical operations (switch a light, block an elevator in case of fire etc.), but also more "ethereal" operations, such as data retention obligations (e.g. "data gathered must be destroyed after 1 month etc."). Here the challenge is more related to the definition, on a side, of a set of languages enabling to express actions, consequences and obligations, and, on the other, of a framework able to translate these obligations into a way that can be understood by all the IoT devices.

5.3.5 Privacy by Design

With the large numbers of IoT devices monitoring the environment, it is almost inevitable that they capture data that can be sensitive to citizens. In one of the previous examples described in this chapter, with the traffic camera, it is obvious that transmitting raw video or still images can potentially breach the privacy of the pedestrians, since their images can be captured by the camera and they can be recognized passing by the street at a specific time. Similarly, with noise measurement devices, which are mainly microphones, conversations between citizens passing by that device can be easily recorded. Furthermore, data transmitted from user devices, even if they are anonymized or pseudonymized, can be easily mapped to individual persons when they are available for a long period of time and if they are linked with information from other sources. For example, when one user sends anonymous location information to an application every day for a long time period, it is easy to extract patterns of movement and when linked to other information it is easy to identify who this person is, where he lives, where he works, etc. One major enemy of "privacy by design" is the reuse of data between applications, because this process allows the linkability of information, which is a main privacy threat. Privacy by design is very much related with the context awareness, since one key mechanism to ensure privacy would be to use context information in order to gather from the device only the exact required information that is needed for a specific application and avoid gathering unneeded data that can raise possible privacy threats [14].

5.3.6 Context Awareness

In order to be able to regulate the interaction of the different sensors and the implementation of operation logics in IoT applications like smart-homes and smart-cities, a way to capture the dynamic evolution of the environment in which the IoT devices are immersed is needed. In other words, the challenge is that of defining a framework able to dynamically modify the behavior of the devices on the basis of the context. This is more relevant under a cyber-security and privacy perspective, as, the same device, in different context, might be required to react in a different manner to address the cyber-security requirements imposed.

Context awareness is actually the topic addressed in this chapter and here we can give a description of the main problems related to this challenge and how to face them. A context based security and privacy framework for IoT has to provide features to dynamically adapt access rules and information granularity to the context. In this book chapter, we use the definition of Context and Context-Aware from [1]. In an emergency crisis scenario for example, private information regarding some possible allergy of a patient should be immediately made available to the doctors but to nobody else, even if the patient cannot give explicit consent in that moment. This means that the context switching should be automatically applied by all the IoT devices involved according to specific security and privacy rules as soon as a change in the context is detected or notified. On the other hand, the system should also be able to avoid malicious users to "emulate" crisis scenarios and impersonate doctors in order to be able to access private user information.

Apart from the detection of a new context, designing this kind of framework presumes a fine definition of the rules and their correlation in the different contexts: the automatics of security and privacy rules defined for a specific context may behave in an incorrect way in a different (or unplanned) context with the consequence of generating vulnerabilities.

Another source of problems can be represented by the sensors/actuators employed by the IoT devices to perform their operations: in normal conditions all the data are collected and processed in a regular way, but for example in a surveillance scenario, sudden worsening of the quality of the images (due to different reasons like hardware failures or malicious activities) may induce false results of the functions implemented in the framework and hamper the overall decision process in the algorithms used to ensure the security and trust of the system [15]. Data integrity in this case is very important since both false positives and misdetections can cause severe problems in the surveillance system. However, in this case, except from the data integrity, the confidentiality of the data should be ensured in a way that the surveillance video should only be disclosed to the administrator of the system and to the persons that have access privileges and not to anyone else.

5.3.7 Summary

In the light of the challenges and problems presented above, the framework we propose is based on the following main concepts:

- security policies implemented as Event-Condition-Action (ECA) enforcement rules;
- specification of Context, Identity, and Role models;
- integrated specification of the IoT System (Structure, Information and Behavior);
- privacy-preserving middleware with behavior-driven services for adaptation to the context;
- secure and privacy-preserving data gathering and transmission at a device level according to security and privacy policies;

- sticky flow policies to annotate a data item in the IoT system and describe how it can be used.

These concepts are explained in more detail in the next sections.

5.4 A Policy-based Framework for Security and Privacy in Internet of Things

The design and implementation of governance and security functions for IoT is done using a Model-based Security Toolkit named SecKit. The SecKit supports Policy Management and Enforcement at all layers of the infrastructure proposed by the iCore project [16] consisting of Virtual Objects (VOs), Composite Virtual Objects (CVOs) and Services. SecKit is based on a collection of meta models, which provide the foundation for security engineering tooling, add-ons, runtime components, and extensions to address requirements of privacy and data protection.

In SecKit, the modeling of the IoT system for security specification purposes is done using a generic design language to represent the architecture of a distributed system across application domains and levels of abstraction inspired into an existing language called ISDL [17]. In SecKit metamodels, the system design is divided into two domains named *entity domain* and *behavior domain*, with an assignment relationship between entities and behaviors. In the entity domain, the designer specifies the entities and interaction points between entities representing communication mechanisms. In the behavior domain the behavior of each entity is detailed including actions, interactions, causality relations, and information attributes.

Using SecKit it is possible to specify, in addition to the system behavior model, the data, identity, context, trust, role, structure, risk, and security rules model. Using these set of metamodels as a reference for the specification of security, trust and privacy rules, our aim is to address the non-functional requirement for interoperability, since these models can be used as a reference for conceptual agreements between different domains running the iCore infrastructure. Figure 5.1 illustrates the SecKit metamodels and their dependencies.

The context model specifies types of Context Information and Context Situations. Context Information is a simple type of information about an entity that is acquired at a particular moment in time, and Context Situations are a complex type that models a specific condition that begins and finishes at specific moments in time [18]. For example, the "GPS location" is a Context

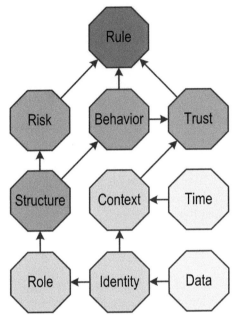

Figure 5.1 SecKit metamodels and dependencies

Information type, while "Fever" and "In One Kilometer Range" are examples of situations where a patient has a temperature above 37 degrees Celsius, or a target entity has a set of nearby entities not further than one kilometer away. Patient and target are the roles of the different entities in that specific situation.

The result of the context situation monitoring are events generated when the situation begins and ends. These events contain references to the entities that participate in the situation and can be used to support the specification of the policy rules. Policy rules can be specified to represent authorizations to be granted when a situation begins and data protection obligations that should be fulfilled when the situation ends. For example, access to the patient data can be allowed when an emergency situation starts with the obligation that all data is deleted when the emergency ends. In another scenario, a security policy may be specified to allow access to data when the situation starts and to trigger the deletion of the data when the situation ends. Existing policy language standards like XACML [19] only support the specification of context as attributes and of textual obligations to be fulfilled when the access to data is granted and not in the future.

The security policies have to be disseminated to the devices that are gathering the data under consideration in a secure way. Depending on the security policy, the device has to trigger and apply the appropriate mechanism for transmitting the data in the exact format needed by the application. This includes a two-step process; at first the device has to map the policies for the application to specific data gathering policies and then it should identify the encryption/security level of the data to identify the proper transmission mechanisms, considering also the energy efficiency requirements of the devices (using i.e. an adaptive encryption scheme as described in [20]). For example, in a traffic monitoring scenario, users in cars may be sending information regarding traffic in an application server. The application should know only how much traffic there is at every street segment. The users' phone has the ability to send various types of traffic related data, i.e. exact location every second, speed every second, direction of movement, etc. If the application wants to estimate the traffic, the related policies should be considered by the devices of the users, so only an average speed per time period and street segment is sent, in order to avoid disclosing the exact location of the user at each point of time (ensuring privacy by design). Actually, intermediate nodes (i.e. the gateway) should also consider these policies and send to the application server only aggregated/average data so that the location of the users will be hidden from the application point of view. Other applications that need to know the exact location of the user (depending on their access control policies) will indeed be identified as such by the devices, which will transmit the exact location (i.e. for a person to track his car if it is stolen).

It is evident, thus, that the transmission of the security policies to the devices is of crucial importance for ensuring the security and privacy of the overall system. The system should be able to identify the integrity of the policies that are sent to the devices, so that unauthorized applications will not gain access to privacy-sensitive data.

The security rules model consists of the security rule templates (a.k.a. policies) specified to be enforced and the configuration rules for these templates. Templates can be specified considering the security and privacy non-functional requirements of confidentiality, data protection, integrity, authorization, and non-repudiation. The security rule templates are Event-Condition-Action rules, with the Action part being an enforcement action of Allowing, Denying, Modifying, or Delaying a VO, CVO, or Service operation. Furthermore, the Action part may also trigger the execution of additional actions to be enforced, or to specify trust management policies to increase/decrease the trust evidence for a specific trust aspect.

The security rule semantics is based on temporal logic and is evaluated using a configurable discrete timestep window of observed events, which in this example is of 2 seconds. Details about the security rule model are described in previously published research papers [21–23]. Examples of security policy rules are provided for our scenario implementation in the following section.

The security policy rules can be delegated from one administrative domain to another when the domains interact and exchange data. For example, when a smart home exchanges data with a smart vehicle, the smart home can exchange the policies that regulate the authorizations and obligations associated to the exchanged data that should be enforced by the smart vehicle. This delegation of sticky flow policies must be supported by trust management mechanisms [24] in order to guarantee or increase the level of assurance with respect to the enforcement of the policy rules by the smart vehicle.

5.4.1 Deployment in a Scenario

The scenario in which we want to show the deployment of the framework is made of three different smart environments: a smart home, a smart office and a smart vehicle. The purpose of this subdivision is to show the different behaviors of the IoT devices when the context changes, according to the policies defined in the framework, and the functioning of the framework itself (application of the rules, interaction with devices). A pictorial description of the scenario is shown in Figure 5.2.

A smart home is an environment that can improve the safety of the citizen and improve the efficiency of house management by providing a variety of functions like remote activating/deactivating power sockets, automatic heating systems or automatic alarm systems. An important goal is usability: the complexity of the different sensors and actuators connected by the house networks or the technical know how about these systems should be made easy for the final user. For example, the setup phase usually is not applicable without technical knowledge and background information and there is the risk to generate a digital divide for special classes of citizens like the elderly people. On the other side, old people want to live on their own, but it is dangerous to be without any care (e.g. medical) for a whole day. Every minute saved in the rescue process after a heart attack or a fall is essential for survival or at least much less painful and much less costly in terms of treatment. To gain medical attendance or at least assisted living it is important to apply an easy-to-use and easy-to-install care system that can fulfil different specific user requirements due to an easy-to-manage personalization process.

However, a smart home is not conceived only for these emergency situations in elderly care but it can provide various smart functions: the remote control of all the household electrical appliances in the house, smart locks that automatically unlock when the owner approaches the door or that can be programmed to give single or regular access to other people under certain conditions, offering of cloud services related to local weather, and the detection of dangerous situations. The smart home used in our examples is more similar to this second description.

A smart office can integrate many of the devices employed in the smart home (again smart locks, weather stations, digital agendas) obviously with different policies and behaviours implemented on them. It can also integrate devices specific for the operations carried on in the office, like multimedia boards, projectors, lab equipment and any connection to the services given in the workplace. One key difference with regards to the smart home is that the office is basically a space shared by various employees. In the home case there is actually no real privacy issue with the data that are gathered by the home devices (meaning that they can be accessed by anyone in the home). However, this is not the same in the office case, where multiple persons are working at the

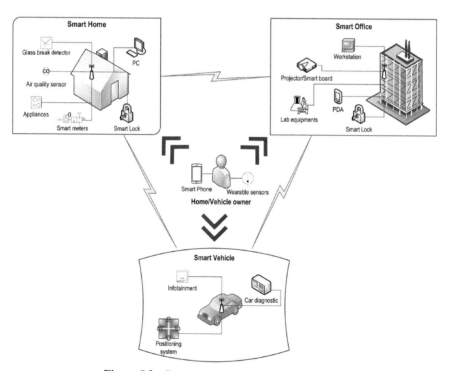

Figure 5.2 Representation of the scenario for IoT

same area and the devices are gathering information for all of them. In this case, the security/privacy policies should not allow the disclosure of sensitive data of one employee to the others, and the applications should enforce those policies on the devices to only gather specific types of data with regards to the end user.

The third environment, the smart vehicle, can be a car able to connect to all the IoT devices carried by the owner or installed at home and in the office. For example, after having checked the presence of the home owner in the car, it can automatically open the gate when it arrives at home, show or send information about the traffic, get information about the working activities when is bringing the owner at the office.

In the next subsection, we will show how a policy implemented in the framework works in the scenario described above, with particular focus on the change of context.

5.4.2 Policies and Context Switching

Figure 5.3 shows the screenshot of the Behavior model section of the SecKit Graphical User Interface (GUI). In this example the Smart City behavior type specifies an interaction (highlighted) between the Smart Vehicle and the Smart Home to Unlock a Smart Lock contained in the Smart Home type.

Figure 5.4 shows the context design model GUI. In this GUI we show the design of the context situations we apply in our policies. We define (i) a situation to detect proximity with a target entity and the set of nearby entities within 20 meters range, (ii) a situation to detect that a person is driving home including the car they are using to drive and the reference to their smart home, and (iii) a situation to detect a health emergency that includes the patient.

Figure 5.5 illustrates trust and context-aware confidentiality policy rules, which are nested, meaning that a combining algorithm must be specified to choose the authorization decision in case both rules are evaluated to be true and are triggered. The outer rule specifies that if the "Access Data" activity is about to be executed by an untrustworthy entity, the decision should be to **Deny** this activity. However, in case of an emergency during the last 3 timesteps, for example in the last 6 seconds for a timestep of 2 seconds, the decision should be to **Allow** the access to the data. For this set of nested rules the combining algorithm chosen is "Allow overrides", meaning that if at least one of the triggered rules in the set allows the activity this decision has priority over any other Deny. The management of trustworthiness values is done in the SecKit using the trust management model proposed by [15].

In addition to allowing or denying access to data it is also possible to specify policies to **Modify** and/or **Delay** the access. A modification could be

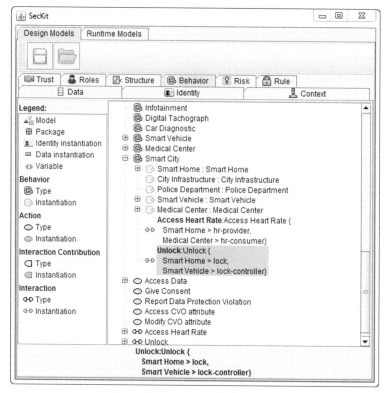

Figure 5.3   Behavior design model

the anonymization of the data access by replacing the identity of the data owner by a pseudonym.

The example in Figure 5.5 shows a policy that can be employed in the scenario proposed focusing on the different behaviour of the components involved when the context changes. For example, let us extend this example to the following situation:

- the home owner is at home watching TV. In this normal condition he has full the control of all the IoT devices installed, showing the presence of some of them used to interact with visitors (e.g. a smart lock to which visitors can ask access) or hiding others that are related to personal activities (agenda, wearable or medical devices);
- vehicles or pedestrians outside can detect the presence of a smart lock that controls the main gate but obviously, if they don't have permission from the owner, they cannot interact with it. This is valid also for emergency

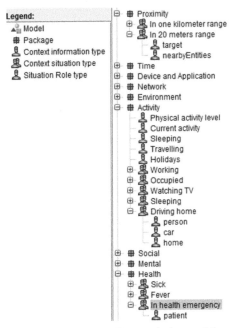

Figure 5.4 Context design model

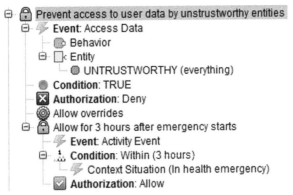

Figure 5.5 Trust and context aware confidentiality policy rule

vehicles or public authorities which in normal conditions are not allowed to enter without the permission of the owner (since they are authorities, recognized by the system, they could be allowed to send a request to the device which would immediately notify the owner, but this can be prohibited to all other people). Considering the device that monitors the lock on the door, it has on board specific policies for being accessed by other users or devices. Cooperating with the overall system in the house, it can also ask for an advice when it receives a request to be opened by a non-authorized person;

- suddenly the health conditions of the owner get worse. He immediately calls the emergency number but he has no time to explain what is happening and faints. Another option is that he presses the "Emergency Button" that is included in many AAL applications. Anyway, an ambulance is sent to his house and the police alerted;
- the IoT devices in the smart home did not receive any information about the context from the owner himself. The only warning is that the telephone has just called an emergency number but no information about people allowed to enter the gate have been modified. In a few minutes the ambulance is already in front of the gate;
- at this point, the medical staff (or the ambulance itself) send opening request to the smart lock controlling the gate. The IoT system in the house does not get any feedback from the owner. However, the system acknowledges that the emergency number has just been called and the sensors worn by the owner detect a lying position and a low heart rate. These conditions immediately trigger a health emergency context and this context (together with the respective policies) is communicated to the devices that are responsible for handling emergency situations. In this respect, the devices that controls the smart lock on the door gets a new policy for allowing access to "Emergency Response Teams", which includes the medical staff and the ambulances. This way, when the ambulance reaches the house, the device on the ambulance has the authorization to access the smart lock and unlock it so that the medical staff can get into the home;
- after the ambulance, the police also arrives and is allowed to interact with the device and give indications about the context. The emergency situation is valid for the duration declared in the policy or until different communication from the owner or police.

The situation described shows how the behaviour of the IoT devices changes when the context changes according to the policies implemented in the framework. Indeed, the crucial point remains the definition of the policies and the detection of the contexts. What if, for example, in this situation the homeowner was simply sleeping? Probably wearable sensors would detect exactly the same activity but the call to the emergency number raises some

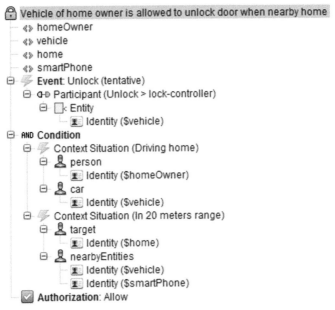

Figure 5.6   Complex context aware access control rule

doubt about this and the gate is open. As an alternative, the owner could define a policy in which medical staff is always allowed to enter.

Another example presented in Figure 5.6, not related to health emergency and in which more smart environments of the same owner are involved, is when the smart vehicle is driving home and it asks for the opening of the gate (tentative Unlock): this request is allowed only if the home owner is actually in the car that triggered the request, if the car is in 20 meters range of the home, and his smart phone is also in the same range. This policy rule template specifies variables for the home owner, vehicle, home, and smart phone.

In this situation, there are some security threats to highlight. If the homeowner is not in the car probably the car has been stolen or another car is trying to enter the gate. If the vehicle is not in front of the gate, it means that someone else, probably malicious who impersonates the home owner or his car, is triggering the opening through the car to gain access to the home. In this case, the system should be capable of realising whether indeed it is the home owner that requests access or not. This can be done, e.g., by accessing other resources that can provide the location of the owner or his habits/patterns/etc. For example, the house system may have access to the office system to check if the home owner is indeed at his work and if so this will mean that an unauthorized person is requesting access to the house. Therefore, the policies have to check all these conditions and try to detect the actual context in order to apply the right behaviour.

The component responsible for the opening of the gate is the smart home, which has to check all the conditions above. For example, it needs proofs that the car is close to the gate (this can be done analysing real time images, showing visual codes like blinking lights or with small range communications encrypted using specific keys) or check the presence of the owner in the car. That could be easy if the owner has its smartphone but what if he forgot it at office? The system could deduce that he is still at work and the car has been stolen. In this case, some cross checks (like comparing the movement of the sensors worn by owner with the movement of the car or prove the identity with some unlocking procedure) can be implemented to solve uncertain situations but, in general, is the definition of the policies that must guarantee a consistent behaviour of the IoT devices. Furthermore, the system should be ensured that the policies are securely sent to the devices, because a malicious user could also send false policies for to get access. For example, one could transmit to the smart lock a policy for an emergency context, impersonating a medical staff so that he can access the lock. The device that controls the lock should be able to identify the trustworthiness of the origin and the integrity of the policy in order to avoid such situations.

All this issues highlight the importance of the context situation detection mechanism and the complexity and level of security required for each home owner requirements and risks. Some home owners may decide to specify additional checks considering the threats of their neighbourhood and the value of their assets at home. All different policies and requirements can be specified using our proposed framework.

5.4.3 Framework Architecture and Enforcement

Figure 5.7 shows the SecKit enforcement components. In our enforcement architecture the IoT Framework and platform are monitored by a technology specific Policy Enforcement Point (PEP), which observes and intercepts service, CVO, and VO invocations taking into account event subscriptions of a Policy Decision Point (PDP). The PEP component signals these events to the PDP, and receives enforcement actions in case a tentative event is signalled. If required for policy evaluation, the PDP may implement custom actions to retrieve status information of VOs and CVOs, and subscribe to context

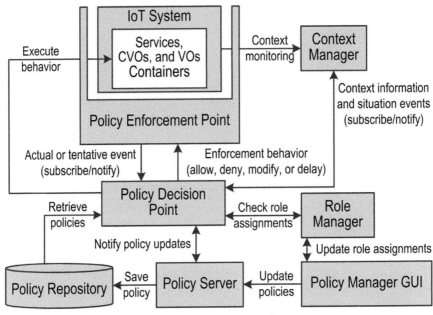

Figure 5.7   SecKit Enforcement Components

information and situation events with the Context Manager component, both using existing functionality provided by the IoT Framework.

In order to be useful in a concrete implementation scenario, the SecKit must be extended with technology specific runtime monitoring components. In the iCore project we provide one extension to support monitoring and enforcement of policies for a MQTT broker, which is the technology adopted by most of the project partners to support communication between VOs, and CVOs. The SecKit may be used in a hospital scenario where VOs and CVOs represent the staff and medical devices being used that communicate using a MQTT middleware. Policies are specified to control access to the hospital staff information (e.g. location) and to control the access to medical devices represented as VOs.

Figure 5.8 shows the runtime interface of the rule engine that instantiates the specified policy rules and receives events generated by extended MQTT broker for a hospital scenario. Our extension is a connector that intercepts the messages exchanged in the broker with a publish-subscribe mechanism, notifies these messages as events in the SecKit, and optionally receives and enforces actions to be executed (e.g. Allow, Deny, Modify, etc.).

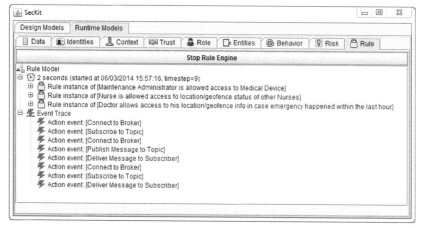

Figure 5.8   MQTT events received by SecKit

5.5 Conclusion and Future Developments

As we already indicated in the introduction, choices that society makes are subject to change, which are based on experiences with technology and the understanding of the issues. The trade-off between security and privacy cannot be determined by technology research alone, it requires societal interaction. However, as research results have shown, technology can enable a better balancing between security and privacy, for instance, by making it possible to limit communications to those parts of data sets that are necessary in the moment. In this book chapter, we proposed an approach based on the SecKit in which policies can be used to control the access and the flow of user's data to address security and privacy. The advantage of this approach is to give the user the control of his own data.

A limitation of the approach presented is that the perception of the context considered does not address potential ambiguity and quality of the data collected by the sensors. This aspect, which is actually based on the fine definition and detection of the context, will be addressed in future developments of the framework. In addition, the adopted scenario has involved a limited number of entities, but in the future IoT each IoT device has to interact with a large number of interfaces. To address this last aspect, related to scalability, we will investigate solutions based on cluster approaches and cloud computing, partitioning of the monitoring function to minimize the flow of data and computation overhead.

5.6 Acknowledgments

This work is partially funded by the EU FP7 Projects REliable, Resilient and secUre IoT for sMart city applications (RERUM) grant agreement n° 609094, and Internet Connected Objects for Reconfigurable Ecosystem (iCore) grant agreement n° 287708.

References

[1] Abowd G. D., Dey A. K., Brown P. J., Davies N., Smith M., and Steggles P., Towards a better understanding of context and context-awareness, in Proc. 1st international symposium on Handheld and Ubiquitous Computing, ser. HUC '99. London, UK: Springer-Verlag, 1999, pp. 304–307.

[2] Embracing the Internet of Everything To Capture Your Share of $14.4 Trillion, Joseph Bradley, Joel Barbier, Doug Handler, CISCO White Paper, 2013.

[3] International Telecommunication Union, "ITU internet reports 2005: The internet of things," International Telecommunication Union, Workshop Report, November 2005.

[4] Guillemin P. and Friess P., Internet of things strategic research roadmap, The Cluster of European Research Projects, Tech. Rep., September 2009.

[5] Miorandi D., Sicari S., Pellegrini F. D., Chlamtac I., Internet of things: Vision, applications and research challenges, Ad Hoc Networks 10 (7) (2012) 1497.

[6] Schindler H.R., Cave J., Robinson N., Horvath V., Hackett P.J., Gunashekar S., Botterman M., Forge S., Graux H.. Europe's policy options for a dynamic and trustworthy development of the Internet of Things, RAND Europe. Prepared for European Commission, DG Communications Networks, Content and Technology (CONNECT).

[7] IoT European Research Cluster. http://www.internet-of-things-research .eu/. Last Accessed 13/May/2015.

[8] Perera, C. Zaslavsky, A. Christen, P. Georgakopoulos, D., "Context Aware Computing for The Internet of Things: A Survey," *Communications Surveys & Tutorials, IEEE*, vol.16, no.1, pp.414,454, First Quarter 2014.

[9] Dey A. K., Abowd G. D., and Salber, D., A conceptual framework and a toolkit for supporting the rapid prototyping of context-aware

applications, Hum.-Comput. Interact., vol. 16, pp. 97–166, December 2001.

[10] Chen H., Finin T., Joshi A., Kagal, L., Perich, F. and Chakraborty D., Intelligent agents meet the semantic web in smart spaces, IEEE Internet Computing, vol. 8, no. 6, pp. 69 – 79, nov.-dec. 2004.

[11] Gessner, D.; Olivereau, A.; Segura, A.S.; Serbanati, A., "Trustworthy Infrastructure Services for a Secure and Privacy-Respecting Internet of Things," Trust, Security and Privacy in Computing and Communications (TrustCom), 2012 IEEE 11th International Conference on, vol., no., pp.998,1003, 25-27 June 2012.

[12] Huang X., Fu R., Chen B., Zhang T., Roscoe, A.W., User interactive Internet of things privacy preserved access control, Internet Technology And Secured Transactions, 2012 International Conference for , vol., no., pp.597,602, 10–12 Dec. 2012

[13] ISO/IEC 24760-1:2011 Information technology—Security techniques—A framework for identity management—Part 1: Terminology and concepts.

[14] Pohls H. et al., "Rerum: Building a reliable iot upon privacy- and security-enabled smart objects," in Proc. of WCNC, 2014.

[15] Neisse, R. Trust and privacy management support for context-aware service platforms. PhD thesis, University of Twente. CTIT Ph.D. Thesis Series No. 11–216 ISBN 978-90-365-3336-2.

[16] Vlacheas, P.; Giaffreda, R.; Stavroulaki, V.; Kelaidonis, D.; Foteinos, V.; Poulios, G.; Demestichas, P.; Somov, A.; Biswas, A.R.; Moessner, K., Enabling smart cities through a cognitive management framework for the internet of things, *IEEE Communications Magazine, ,* vol.51, no.6, pp.102,111, June 2013.

[17] Quartel D., Action relations-basic design concepts for behaviour modelling and refinement, PhD Thesis University of Twente, 1998.

[18] Pereira, I. Dockhorn Costa, P. Almeida, J. P. A. A Rule Based Platform for Situation Management. In: 2013 IEEE International Multi-Disciplinary Conference on Cognitive Methods in Situation Awareness and Decision.

[19] Rissanen A., eXtensible Access Control Markup Language v3.0, Available at: http://docs.oasis-open.org (2010).

[20] Fragkiadakis A., Charalampidis P., Tragos E., Adaptive compressive sensing for energy efficient smart objects in IoT applications, Wireless Vitae 2014, (accepted, to appear).

[21] Neisse, R. Pretschner, A. Di Giacomo, V. A Trustworthy Usage Control Enforcement Framework. The International Journal of Mobile

Computing and Multimedia Communications (IJMCMC), 5(3), 34–49, July-September 2013.

[22] Neisse, R. Doerr, Joerg. Model-based Specification and Refinement of Usage Control Policies. Eleventh International Conference on Privacy, Security and Trust (PST), Tarragona, Spain, Jul. 2013.

[23] Neisse, R. Pretschner, A. Di Giacomo, V. A Trustworthy Usage Control Enforcement Framework. Proc. 6th Intl. Conf. on Availability, Reliability and Security (ARES), Vienna, Austria, Aug. 2011.

[24] Neisse, R. Holling, D. Pretschner, A. Implementing Trust in Cloud Infrastructures. 11th IEEE/ACM International Symposium on Cluster, Cloud and Grid Computing (CCGRID), Newport Beach, USA, May 2011.

6

Scalable Integration Framework for Heterogeneous Smart Objects, Applications and Services

Sébastien Ziegler,[1] Maria Rita Palattella,[2] Latif Ladid,[2]
Srdjan Krco[3] and Antonio Skarmeta,[4]

[1]*Mandat International, Switzerland*
[2]*University of Luxembourg, Luxembourg*
[3]*Ericson, Serbia*
[4]*University of Murcia, Spain*

6.1 Introduction

Over the last decades, the Internet has had a profound effect on the way we live and conduct business. The original ARPANET was conceived as a simple and reliable network of interconnected servers but the standardization of TCP/IP [1–2] between 1974 and 1982 has unexpectedly paved the way to the largest single market of human history. Since the 90s, the Web has emerged and encompassed a huge numbers of connected applications and services. As more and more systems and actors were connected to the Internet the emergence of digital and social platforms was still a rather natural development, using the very same Internet architecture.

For years, there was an implicit expectation that the growth of the Internet would be limited in a way which correlates to the World population. This expectation was continually strained as the number of web sites and users connected to the Internet continued to grow and is not valid anymore, as we have entered a new era, namely the Internet of Things' Era. We are moving beyond a point of no return, with already more devices connected to the Internet than human beings (there will be over 50 Billion connected devices by the end of this decade). Every day the devices are becoming smarter, more pervasive and more mobile. The Internet is already used as a vehicle for many Machine to Machine (M2M) connections, as it is used for Voice over IP and EPC tags management. Actually, the Internet is progressively becoming a broad platform for the connectivity of many kinds of entities. Among them, machine-to-machine and machine-to-human communications will be more numerous than human initiated activities.

6.2 IPv6 Potential

Since 1982, the Internet has benefited from the stable and well-designed Internet Protocol version 4 (IPv4) [1]. However, IPv4 only has a capacity of about 4 billion theoretical public addresses (and fewer in practice). This corresponds to less than one public IP address per living adult on Earth – a number that was believed to be sufficient to address current and future needs at the time of its creation. Progressively, however, the growing allocation of public Internet addresses started to cause concerns, leading to restricted public allocation policies and the introduction of Network Address Translation (NAT) mechanisms to provide end-users with private addresses. Most users effectively became "Internet homeless", unaware that they were sharing potentially temporary public Internet addresses with others.

The opening of the Internet for commercial use and its growth prompted the IETF to design a new protocol with a larger addressing scheme, standardized in 1998 as the Internet Protocol version 6 (IPv6) [3]. The IPv6 protocol is based on an addressing scheme of 2^{128} bits, split in two parts: 2^{64} bits for the network address and 2^{64} bits for the host ID. IPv6 is now globally deployed [4] and a growing number of Internet Service Providers (ISP) is offering IPv6 connectivity.

The extended scheme offered by IPv6 enables a virtually unlimited number of addresses, overcoming the scarcity issues of IPv4 and catering thereby for the exploding needs of the Internet of Things. The addressing scheme now available provides the possibility to allocate unique public Internet addresses to as many devices as needed, making each and every smart object Internet accessible through a unique IPv6 address.

IPv6 is emerging as the natural answer to the emerging Internet of Things requirements. It provides a highly scalable addressing scheme as well as many useful features (e.g., stateless configuration mechanisms) and a native integration to the future Internet.

In parallel to IPv6, several IPv6-related standards have emerged, including among others: the IPv6 over Low power WPAN (6LoWPAN) [5] providing a lighter version of IPv6 for constrained nodes and networks; the IPv6 Routing Protocol for Low-Power and Lossy Networks (RPL) [6]; the Constrained Application Protocol (CoAP) [7] providing a light substitute to HTTP; the Network Mobility protocol (NEMO) [8], providing mobility support for entire networks of IP devices. Still, new Working Groups (WGs) have been created at IETF, in order to develop others IPv6-enabled protocols. For instance, the newly formed 6TiSCH WG [9] aims to link the IEEE802.15.4e Time Slotted Channel Hopping (TSCH) MAC with IPv6 (and in detail, with 6LoWPAN and RPL).

6.3 IoT6

IoT6, a 3 years (2011–2014) FP7 European research project [10], aimed at exploiting the potential of IPv6 and related standards (6LoWPAN, CoAP, etc.) to overcome current shortcomings and fragmentation of the Internet of Things, in line with the Internet of Things European Research Cluster (IERC) vision and the EC recommendations. Its main challenges and objectives were to:

1. Research the potential of IPv6 features and related standards to support the future Internet of Things and to overcome its current fragmentation.
2. Design an Open Service Layer to provide mechanisms for discovery, look-up and integration of services offered by Smart Objects to distributed clients and devices connected via IPv6.
3. Explore, based on Service-Oriented Architecture, innovative forms of interactions with:
   - Information and intelligence distribution;
   - Multi-protocol interoperability with and among heterogeneous devices, including various non-IP based communication protocols;
   - Device mobility and mobile phone networks integration;
   - Cloud computing integration with Software as a Service (SaaS);
   - Tags and Smart Things Information Services (STIS) [11].

In other words, IoT6 has explored the potential of IPv6 for horizontal integration (across various domains of the IoT) and vertical integration between the IoT and the Cloud. The main outcomes of the IoT6 project are recommendations on how IPv6 features can be exploited for accelerating the development of the Internet of Things, together with a well-defined IPv6-based Service Oriented Architecture enabling interoperability, mobility, cloud computing and intelligence distribution among heterogeneous smart components, applications and services, including business processes management tools and smart buildings.

IoT6 has demonstrated the high potential of IPv6 for the future IoT, by providing an ideal solution to interconnect unlimited number of heterogeneous smart things, as well as a powerful integrator for the integration of the Internet of Things with Cloud applications and web services. IoT6 has worked in close cooperation with International Forums (e.g., IPv6 Forum, ITU-T JCA-IoT), standardization bodies (e.g., ETSI, M2M, ETSI, IETF), major industries and other research projects (e.g., IoT-A, IoT-I, SEnsei, etc.) with a European and international perspective.

6.4 IPv6 for IoT

Why should the Internet of Things care about IPv6? Many answers can be given to such question, and thus, there are several arguments that show IPv6 will be (and actually it is already) a key enabler for the future Internet of Things:

1) *Adoption is just a matter of time*

The Internet Protocol is a must and a requirement for any Internet connection. It is the addressing scheme for any data transfer on the web. The limited size of its predecessor, IPv4, has made the transition to IPv6 unavoidable. The Google's figures are revealing an IPv6 adoption rate following an exponential curve, doubling every 9 months about [4].

2) *Scalability*

IPv6 offers a highly scalable address scheme. It provides 2^{128} unique addresses, which represents 3.4×10^{38} addresses. In other words, more than 2 Billions of Billions addresses per square millimetre of the Earth surface. It is quite sufficient to address the needs of any present and future communicating device.

3) *Solving the NAT barrier*

Due to the limits of the IPv4 address space, the current Internet had to adopt a trick to face its unplanned expansion: the Network Address Translation (NAT). It enables several users and devices to share the same public IP address. This solution is working but with two main trades-off:

- The NAT users are borrowing and sharing IP addresses with others. Hence, they do not have their own public IP address, which turns them into homeless Internet users. They can access the Internet, but they cannot be directly accessed from the Internet.

- It breaks the original end-to-end connection and dramatically weakens any authentication process.

4) *Strong Security enablers*

IPv6 provides end-to-end connectivity, with a more distributed routing mechanism. Moreover IPv6 is supported by a very large community of users and researchers supporting an on-going improvement of its security features, including IPSec.

5) *Tiny stacks available*

IPv6 application to the Internet of Things has been being researched since many years. The research community has developed a compressed version of IPv6 named 6LoWPAN. It is a simple and efficient mechanism to shorten the IPv6 address size for constrained devices, while border routers can translate those compressed addresses into regular IPv6 addresses. In parallel, tiny stacks have been developed, such as Contiki, which takes no more than 11.5 Kbyte.

6) *Enabling the extension of the Internet to the web of things*

Thanks to its large address space, IPv6 enables the extension of the Internet to any device and service. Experiments have demonstrated the successful use of IPv6 addresses to large scale deployments of sensors in smart buildings, smart cities and even with cattle. Moreover, the CoAP protocol enables the constrained devices to behave as web services easily accessible and fully compliant with REST architecture.

7) *Mobility*

IPv6 provides strong features and solutions to support mobility of end-nodes, as well as mobility of the routing nodes of the network.

8) *Address self-configuration*

IPv6 provides an address self-configuration mechanism (Stateless mechanism). The nodes can define their addresses in very autonomous manner. This enables to reduce drastically the configuration effort and cost.

9) *Fully Internet compliant*

IPv6 is fully Internet compliant. In other words, it is possible to use a global network to develop one's own network of smart things or to interconnect one's own smart things with the rest of the World.

6.5 Adapting IPv6 to IoT Requirements

The IoT requires software architectures that are able to deal with a large amount of information, queries, and computation, making use of new data processing paradigms, stream processing, filtering, aggregation and data mining. In a regular Internet environment, this is sustained by communication standards such as HyperText Transfer Protocol (HTTP) [12] and Internet Protocol (IP) [1].

In contrast, some IoT objects are requiring very low power consumptions in order to be powered by batteries or through energy-harvesting. Energy is wasted by the transmission of unneeded data, protocol overhead, and non-optimized communication patterns; these need to be taken into account when plugging objects into the Internet. Existing Internet protocols such as HTTP [12] and Transmission Control Protocol (TCP) [2] are not optimized for very low-power communication, due to both verbose meta-data and headers, and the requirements for reliability through packet acknowledgement at higher layers, which hinders the adaptation of existing protocols to run over that type of networks. In order to interconnect as well as Internet-connect several IoT devices (e.g., RFID, sensors, machines, etc.), a low power, highly reliable, and Internet-enabled communication stack is needed [13].

Aware of that, IoT6 has adopted a protocol stack including IEEE802.15.4 PHY-MAC, 6LoWPAN, RPL, and CoAP, and thus able to fulfil the requirements of constrained devices. In detail, IoT6 devices are based on the 6LoWPAN protocol, backed by IEEE802.15.4 gateways. Within small IPv6 clusters, the resource and service discovery has been performed using the Multicast DNS (mDNS) [14] and Resource Directory (RD) functionality, combined together. Instead, within large IPv6 clusters, the resources have been connected to the global discover engine based on DNS-Service Discovery (i.e., DigCovery) [15].

6.6 IoT6 Architecture

The IoT6 architecture has been designed by taking into account to the furthest possible extent the outcomes of other relevant projects, most notably IoT-A (i.e., the IoT ARM [16]), ETSI M2M [17] and FI-WARE [18]. These outcomes were adapted and enhanced with IoT6 specific features and components, mainly coming from project's reliance on IPv6 functionality. The aim was to utilize the properties of this protocol and to re-use them within the architecture model, possibly replacing some of the standard components. For example, parts of the service and resource discovery functionality has been replaced with the DNS-SD [15] and mDNS [14] based approaches. As shown in the IoT ARM Functional Model in Figure 6.1, IoT6 has contributed mainly to the Communication, Service organization, IoT service and Security components.

The initial IoT6 architecture design approach followed the initial IoT ARM Guidances that were available at that time. Then, it was mainly relying on modification of already available ETSI M2M and FI-WARE IoT architectures. The resulting IoT6 architecture is shown in Figure 6.1.

On the device level, it is possible to distinguish devices supporting IPv6, and legacy devices (i.e., devices not supporting it). IPv6-based devices can be organized in small or large clusters. Legacy devices can support a range of specific protocols, such as KNX [19], ZigBee [20], or Bluetooth [21], as well as IPv4. An additional cluster is dedicated to EPC global compliant RFID system.

At the communication level, IoT6 utilizes IPv6 (and 6LoWPAN for low power devices). Devices are connected either via the so-called half gateways (that convert legacy protocols to IPv6) or directly, when they are IPv6-enabled. This setup can be directly mapped to the IoT ARM communication channel model [16]; IoT ARM's constrained networks are mapped to one or the other group of devices as defined above, while IoT6's half-gateways represent IoT ARM's gateways. On top of the IPv6 layer, CoAP has been selected as the preferred protocol with different encoding techniques (JSON, XML). For a specific case of building automation, oBix protocol was also included.

At the IoT service level, the IoT6 architecture support several solutions. In the case of small IPv6 clusters, mDNS is used for service registration and discovery (inside the cluster). In the case of large clusters, DNS-SD is used for internal cluster service registration and discovery. For the EPCIS cluster, an adaptation of the Digcovery solution was needed. On the global level, two solutions are supported: Digcovery (see Sec. 6.7) and CoAP Resource Discovery (RD). When it comes to the service organization level, the project relies on the cloud based workflow and process management services which interact with the rest of the system using CoAP.

6.7 DigCovery

An important outcome of the IoT6 project is the DigCovery platform, shown in Figure 6.2, and composed by a DigCovery and a DigRectory. DigRectory consists in an independent local resource directory that collects services provided by smart devices such as RFID cards, legacy devices and 6LoWPAN devices. These digrectories are managed through DNS-queries extended with a

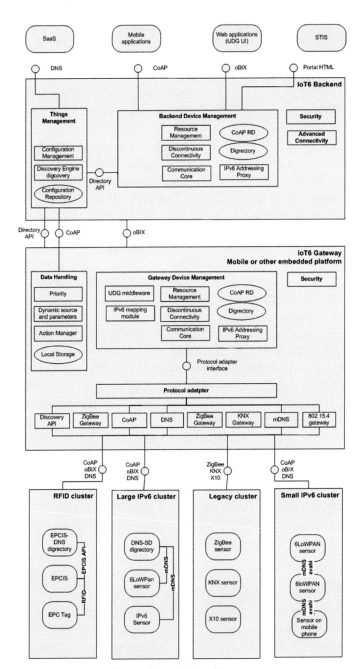

Figure 6.1 IoT6 Architecture

Figure 6.2 DigCovery Platform

search engine. In order to make the system scalable, it offers a centralized point, called DigCovery core, to manage and discover digrectories. The DigCovery platform components can be grouped into 3 classes. The low level corresponds to local discovery. DigRectory is responsible to detect any sensor with a service discovery protocol announcement. The mDNS and CoAP protocols are supported as service announcement protocol. The mid-level corresponds to DigCovery. DigCovery is responsible to make public a private service that is stored in a local DigRectory. For this reason, such level is called global discovery. The top level corresponds to DigCovery protocols and applications developed for DigCovery.

6.8 IoT6 Integration with the Cloud and EPICS

IoT6 architecture has been designed to enable direct integration of the Internet of Things with the cloud. The IoT6 stack has been deployed on Software as a Service platform enabling direct interaction between Cloud-based services and locally deployed sensors and actuators. The use of CoAP appeared to be well suited for such integration, enabling large scale deployments and a direct and REST compliant interaction between the services and the smart things, paving the way to a large scale Web of things. In parallel, the IoT6 platform has explored the integration and interaction with smart things information systems, such as EPICS, traditionally to manage RFID or other similar tags. The experiments made so far have enabled to extend the use of such systems to sensors and more complex devices. It also enables the EPICS to interact with IPv6-enabled things, regardless of its location and without requiring a reader in between [22].

6.9 Enabling Heterogeneous Integration

One of the challenges for the future Internet of Things is related to its inherent heterogeneity. Hundreds of communication protocols have emerged to address specific requirements. Interconnection of things implies to deal with huge amount of different technologies and then with their different protocols. Some technologies were developed with IP capabilities; others used different networking technologies, with open or proprietary buses. Over time, part of those protocols may move towards IP. However, existing systems are likely to remain and quite a number of communication protocols will keep their specific bus technology. The integration of heterogeneous Internet of Things components faces several challenges, including:

- Integrating non-IP-based communication protocols into an IP-based environment
- Integrating together communication protocols using different application layers.

Along the time, different solutions have been researched and developed:

Bridges and gateways
The first and most natural integration scheme has been to develop bridges and gateways enabling the translation of a communication protocol into another one. It enables the integration of distinct protocols into IPv6 and vice-versa. Such gateways usually provide a clear IP-based API to communicate with the devices and its specific communication protocol.

IP adaptation
Several communication protocols have moved a step farther by developing IP-based versions of their own protocols. This option has been largely developed in the building automation domain, with protocols such as the KNX Association, which has standardized a KNX IP version of its standard.

Universal Device Gateway

The Universal Device Gateway (UDG) [23] is a multi-protocol control and monitoring system developed by a research project initiated in Switzerland in 2006. It aimed at integrating heterogeneous communication protocols into IPv6. The UDG control and monitoring system enables cross protocol interoperability. It demonstrated the potential of IPv6 to support the integration among various communication protocols and devices, such as KNX, X10, ZigBee, GSM/GPRS, Bluetooth, and RFID tags. It provides connected device with a unique IPv6 address that serves as unique identifier for that object, regardless its native communication protocol. It has been used in several research projects, including in the framework of IoT6, where it has been used among other as an IPv6 and CoAP proxy for all kinds of devices.

IoT6 stack and IoTSyS

IoT6 has designed and tested a protocol suite enabling the integration of various communication protocols into the common IoT6 architecture. It is based on IPv6 (or 6LoWPAN in constrained networks), CoAP, JSON and oBIX. In order to test the integration of legacy protocols, the IoT6 research project has developed IoTSys [24], a prototype of a Java based integration middleware abstracting the low level protocol details through the IoT6 stack to allow the communications with the other components of the IoT6 framework and vice-versa. This prototype was used to test and demonstrate the integration of IoT6 with several protocols such as BACnet, KNX, ZigBee, etc.

IoT6 has confirmed the capacity of those various approaches to integrate heterogeneous communication protocols and devices together through IPv6. While traditional approaches require multiplying the number of bridges for each couple of communication protocols, the two latter solutions enable a simplification of the network extension to additional standards. Moreover, they are easily portable and deployable in constrained environments.

IPv6 Address mapping

Beyond the interconnection and interoperability mechanisms, another issue has been addressed by IoT6: the possibility to map IPv6 addresses on top of other addressing schemes, from non-IP communication protocols. Part of the challenge of integrating legacy technologies into an IPv6 network is represented by devising a mechanism for stateless auto configuration of such devices. Indeed such mechanism would ensure that a number of properties of the mapping hold, such as:

- Consistency: a host should get the same IPv6 address every time it connects to a same legacy network. This feature might be particularly important for devices which are not always "on", or which are not permanently connected

- Local Uniqueness: for devices which have an IPv6 address with a same network part, the host part should be unique for each host. This property avoids address's conflicts within a same subnet.
- Uniqueness within the whole Internet: coherently with the IoT vision, the host part of an IPv6 address associated to a host should be unique within the whole Internet.

This effort within the IoT6 project has produced a proposal for a new standard for IPv6 address mapping of non-IP-based communication protocols, currently in the form of an IETF draft. The proposed solution named 6TONon-IP provides a clear specification of a mapping mechanism which tries to maximize the satisfaction of the properties mentioned. The gateways, once provided through the IPv6 address mapping solution the IPv6 addresses to the objects they manage, must use a semantic to identify and differentiate the protocols. Two solutions were deployed to address this challenge and each one designed its own internal semantics.

6.10 IoT6 Smart Office Use-case

In the context of the IoT6 project several use cases were developed. Among them, hereafter we present the *Smart Office* use-case which demonstrates the ability of the IoT6 architecture to interact with heterogeneous devices, including non-IP based protocols, with a focus on energy efficiency and user comfort. In this use case, an employee arrives at his office building. He identifies himself with a mobile phone through an interface, such as NFC. A terminal reads the tag included in the mobile phone. The lights, the windows and the HVAC system are adapted to create a comfortable and welcoming ambiance for him at his work station. The employee updates custom preferences through the smart phone. The service which manages the communications on the smart phone network does a request to the resource directory to find out the IPv6 address of the local CMS. A visitor arrives and is guided to the waiting lounge. A presence sensor installed in the waiting lounge detects the arrival of the visitor and advises local CMS which starts the video and the music and adapts the lighting. Later, when the employee exits the office, the lights, the windows and the HVAC systems around his work station are automatically adapted in order to save energy. All those interactions are enabled through IPv6 and IoT6 architecture. The IoT Context View of the

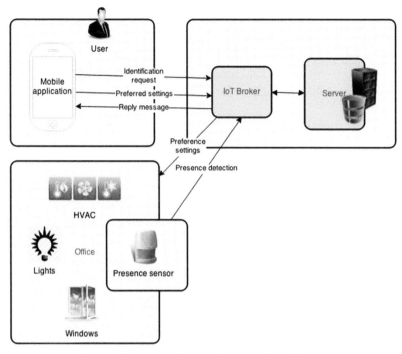

Figure 6.3 Smart Office IoT Context View Diagram

use-case is presented in Figure 6.3 and the corresponding Functional View in Figure 6.4.

6.11 Scalability Perspective

As previously mentioned, an important quality of IPv6 comes from its large addressing scheme able to cope with very high scalability requirements. However, it was important to explore and test the IoT6 scalability from a systemic perspective. In order to do so we have adopted several approaches.

To demonstrate the scalability of IPv6 with real deployment, the IoT6 architecture has been successfully interconnected with all the remotely accessible sensors from the smart city of Santanders. In order to achieve the integration, a UDG has been used as an IPv6 and IoT6 proxy for the Santanders sensors, turning each Santander's sensor into an IoT6 enabled one.

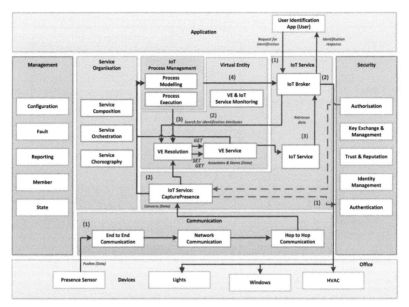

Figure 6.4 IoT6 Functional View

Scalability tests have been performed on individual components of the IoT6 architecture too. For instance, several tests have been carried out in order to validate the capabilities and scalability of the DigCovery System. In detail, such test procedures can be grouped into three groups: Announcement, Registration and Service Search. The Announcement process that consists in making public a service that is available in a smart object (without prior knowledge of its existence) can be done via mDNS or CoAP. The Registration process is carried out by DigCovery Communication Protocol. A registration request is the process through which a DigRectory inserts or updates the domain that represents on the global DigCovery server. This process is expensive, but it can be managed by DigRectory. DigRectory can make a registration request to DigCovery. DigCovery answers depending on whether the request can be attended or not. If not, DigRectory waits and tries it again. By doing so the server does not crash due to a high number of registration requests. Finally, the third and most important test is Service Search. It is necessary to provide information in scalable way to all clients from DigCovery (Global Server). Registering and announcement are eventual operations and are performed from or to DigRectories. Service queries can be done to the local server in order to receive the local domain services. From the scalability tests that have been carried on, it resulted that Digcovery is a scalable system, if it is properly used.

The IoT6 deployment is distributed across Europe and Asia, including among others: Switzerland, France, Spain, UK, Serbia, Vienna and Korea. Most use case are voluntarily adopting a distributed approach to test and check the reliability of the designed architecture in real conditions, including real Internet network infrastructure. Additional experiments have been performed to successfully interconnect the Geneva testbed with Beijing University of Post and Telecommunication.

6.12 Conclusions

During two and a half years of research and experiments, IoT6 has demonstrated that:

- IPv6 provides a reliable solution to address the scalability requirements of the IoT in terms of number of nodes and geographic scope;
- The Internet of Things is likely to keep a certain level of heterogeneity, including several communication protocols - but efficient solutions exist to integrate this heterogeneity into IPv6;
- IPv6 provides many additional features which are relevant for the IoT, such as multicast, anycast, address self-configuration, etc.
- A whole set of complementary standards are being provided to address the specific IoT constrained devices requirements in an IPv6 framework, including 6LoWPAN, CoAP, RPL, NEMO and 6TiSCH;
- IPv6 constitutes a very good candidate to integrate a globally distributed Internet of Things with cloud applications and resources.

Finally, based on our research, we foresee and can anticipate, without taking too much risk, an increasing convergence between IPv6 and the Internet of Things.

References

[1] Postel J., Internet Protocol, RFC 791, Internet Engineering Task Force RFC791, September 1981.
[2] Postel J., Transmission Control Protocol, RFC 793, Internet Engineering Task Force RFC 793, September 1981. [Online]. Available: http://www.rfc-editor.org/rfc/rfc793.txt

[3] Deering S., and Hinden R,, "RFC2460: Internet Protocol, version 6 (IPv6)," 1998, http://tools.ietf.org/html/rfc2460.

[4] IPv6 Traffic and Mobile Networks Stats by Cisco, online at http://6lab.cisco.com /stats/

[5] Kushalnagar N., Montenegro G., and Schumacher C., "IPv6 over Low-Power Wireless Personal Area Networks (6LoWPANs): Overview, Assumptions, Problem Statement, and Goals, RFC 4919", Internet Engineering Task Force RFC 4919, August 2007.

[6] Winter T., Thubert P., Brandt A., Hui J., Kelsey R., Levis P., Pister K., Struik R., Vasseur J. P., and Alexander R., "RPL: IPv6 Routing Protocol for Low-Power and Lossy Networks, RFC 6550", Internet Engineering Task Force RFC 6550, March 2012.

[7] Shelby Z., Hartke K., Bormann C., and Frank B., "Constrained Application Protocol (CoAP)", IETF CoRE Working Group, February 2011.

[8] Devarapalli, V., Wakikawa R., Petrescu A., Thubert P., "RFC 3963: Network Mobility (NEMO) Basic Support Protocol," IETF, NEMO Working Group, January, 2005. http://www.ietf.org/rfc/rfc3963.txt

[9] IETF, "6TiSCH: IPv6 over the TSCH mode of IEEE 802.15.4e", http://datatracker.ietf.org/wg/6tisch/.

[10] IoT6 research project website: http://www.iot6.eu

[11] Giang N. K., Kim S., Kim D., Jung M., and Kastner W., "Extending the EPICS with Building Automation Systems: a New Information System for the Internet of Things," in International Workshop on Extending Seamlessly to the Internet of Things (esIoT, Birmingham, UK), 2014.

[12] Fielding R., Gettys J., Mogul J., Frystyk H., Masinter L., Leach P., and Berners-LeeT., "HyperText Transfer Protocol – HTTP/1.1, RFC2616", Internet Engineering Task Force RFC 2616, June 1999. Available at: http://www.rfc-editor.org/rfc/rfc2616.txt

[13] Palattella M. R., Accettura N., Vilajosana X., Watteyne T., Grieco L. A., Boggia G., and Dohler M., "Standardized Protocol Stack For The Internet Of (Important) Things", IEEE Communications Surveys and Tutorials, vol 15, no 3, 1389–1406 , Third Quarter 2013

[14] Cheshire S., and Krochmal M., "Multicast DNS (mDNS), RFC6762", Internet Engineering Task Force, Febr. 2013

[15] DNS-SD web page: http://www.dns-sd.org/

[16] Intranet of Things-Architecture (IoT-A) EU F7 project web site: http://www.iot-a.eu/public

[17] ETSI-M2M web site: http://www.etsi.org/technologies-clusters /technologies/m2m

[18] FI-WARE web site: http://www.fi-ware.org/

[19] KNX web site: http://www.knx.org/knx-en/knx/association/introduction /index.php

[20] ZigBee Alliance web site: http://www.zigbee.org/

[21] Bluetooth technology website: http:// www.bluetooth.com/Pages/ Bluetooth-Home.aspx

[22] Giang N. K., Kim S., Kim D., Jung M., and Kastner W., "Extending the EPICS with Building Automation Systems: a New Information System for the Internet of Things," in *International Workshop on Extending Seamlessly to the Internet of Things (esIoT, Birmingham, UK)*, 2014.

[23] Universal Device Gateway website: http://www.devicegateway.com.

[24] IoTSys available at : http://code.google.com/p/iotsys

7

Internet of Things Applications - From Research and Innovation to Market Deployment

Maurizio Spirito,[1] Claudio Pastrone,[1] John Soldatos,[2] Raffaele Giaffreda,[3] Charalampos Doukas,[3] Vera Stavroulaki,[4] Luis Muñoz,[5] Veronica Gutierrez Polidura,[6] Sergio Gusmeroli,[7] June Sola,[7] Carlos Agostinho,[8]

[1]*ISMB, Italy*
[2]*Athens Information Technology, Greece*
[3]*CREATE-NET, Italy*
[4] *UPRC, Greece*
[5]*Universidad de Cantabria, Spain*
[6]*TXT e-solutions, Italy*
[7]*Innovalia Association, Spain*
[8]*UNINOVA, Portugal*

7.1 Introduction

The IoT has received considerable and growing attention in the recent years due to its potential to radically change our daily lives. There is almost no application domain where IoT cannot find an application and, most of all, there is no application domain where IoT does not have disruptive potentials. This has generated a lot of expectations for the uptake of IoT-based solutions.

The biggest challenge that needs to be faced when shifting research and innovation results to the market is to overcome the barriers generated by the fragmentation of IoT, both in terms of technologies and systems (e.g., Cloud technologies, Big data, cyber-physical systems, network technologies, privacy & security technologies) and in terms of application domains (e.g., e-health, energy efficiency, smart grids, intelligent transport systems, environmental monitoring and logistics, etc.).

The European Commission has put a huge effort in stimulating collaboration between stakeholders from different domains and in fostering joint research and innovation projects with the goal in mind of creating the multistakeholder ecosystem that is key for the success of IoT. This chapter tries to give a glimpse of where this process stands by describing some of these EU-funded research and innovation projects.

It is impossible to provide a comprehensive overview of the overall situation in Europe; in fact, the selected projects provide a view - not meant to be exhaustive - on how they have planned and executed their activities and what strategies they have adopted to maximise impacts and ensure take up of their solutions. Four stages, in fact, can be generally identified in a project lifetime: (i) Design (before the project starts), (ii) Execution (during the project), (iii) Results (when the project ends), (iv) Acceptance and Sustainability (after the project ends). Depending on the maturity of the projects identified (i.e., whether they are closer to the Design or to the Result stage at the time of writing this chapter), different snapshots of the lifetime are provided. For instance, in the Design stage, the key issues addressed relate to what a project plans to implement or demonstrate. They also relate to which proof-of-concepts are conceived and why. Already in the Design stage it is important to devise concrete actions to ensure take up of the final solutions delivered at the end of the project. During the execution stage usually unforeseen issues arise (e.g., in the deployments) that need to be addressed as well as new opportunities for maximising impacts. In these circumstances, projects adopt countermeasures or adjust their plans. At the end of their execution, projects are able to summarize their achievements, compare them with the results that were expected when the project was designed and, most of all address the lessons learned during project execution. These relate, for instance, with acceptance of the solutions created and with the potential for exploitation (e.g., IPRs generated, new pre-commercial prototypes, new businesses identified, etc.). The fourth and last stage, after the project ends, is the most important for commercial uptake. Success happens only if a solid sustainability plan is implemented to overcome the barriers created by the multi-stakeholder nature of IoT and only if convincing acceptance measures are put in place to win resistance to the potentially disruptive impact of IoT on consolidated social, economic and production processes.

The above aspects have inspired the sections that follow, which have been contributed by key Representatives of the following EU-funded projects: OpenIoT, iCore, COMPOSE, SmartSantander, FITMAN, OSMOSE.

7.2 OpenIoT

Cloud computing [1] and Internet-of-Things [2] are two of the main pillars of the Future Internet. Few years after the introduction of these two novel computing paradigms, it became apparent that significant benefits could emerge from their convergence.

7.2.1 Project Design and Implementation

The OpenIoT project (incepted/proposed in 2010) was highly motivated by the need for effectively blending IoT and cloud computing concepts [3]. At that time, several efforts towards IoT/cloud integration had been undertaken both in the research community (e.g., [4], [5], [6]), but also in enterprise world (e.g., Xively (xively.com) formerly known as Pachube). A common characteristic of these efforts was their ability to stream IoT data to the cloud in order to benefit from its scalability and capacity. However, all of these efforts were characterized by prominent limitations and weaknesses, which OpenIoT proposed to remedy.

One of these limitations was the essential lack of (semantic) interoperability [8] between different IoT deployments. One of the main goals of OpenIoT was to unify the semantics of different IoT deployments in the cloud. To this end, OpenIoT proposed the use of the W3C Semantic Sensor Networks (SSN) ontology [9] as a standards-based common model for semantic unification of diverse IoT systems and data streams. Apart from adhering to the W3C SSN standards for modelling and representing sensors and IoT data streams, OpenIoT was also designed to exploit other semantic web technologies such as the Linked Data concept towards linking related sensor data sets.

In terms of technological design and implementation, OpenIoT was also motivated from background developments of the partners such as the popular Global Sensor Networks (GSN) middleware [10], which enables the streaming and integration of diverse sensors and Wireless Sensor Networks based on nearly zero programming. However, at the time of OpenIoT inception, GSN was still using a simple mainstream RDBMS (Relational Database Management System) for persisting and managing data, which was associated with significant limitations. Therefore, cloud integration was deemed as a natural step to the evolution of the GSN open source middleware.

Another novel aspect of OpenIoT concerned the successful blending of several cloud computing concepts (e.g., the on-demand utility-based pay-as-you-go access to resource) into OpenIoT applications. In particular, OpenIoT was designed to support on demand access to available IoT resources in the cloud, thereby enabling a novel utility-based model for IoT such as «Sensing-as-a-Service» applications. Furthermore, it was planned that OpenIoT would be an open source infrastructure for IoT/cloud integration. The consortium believed that an open source project could become a vehicle for wide adoption of the project's results, within both research and enterprise communities. OpenIoT aspired to become a popular open source middleware for IoT/cloud integration, which could be used extensively for research and education purposes, and possibly (following some additional development and fine-tuning) for enterprise developments. Overall, OpenIoT was planned as a joint effort of prominent open source contributors towards enabling a new range of open large scale intelligent IoT applications according to a utility computing delivery model.

7.2.2 Execution and Implementation Issues

The research, design and development of the OpenIoT infrastructure were associated with various implementation and deployment challenges. Several of these challenges stemmed from the need to successfully blend and integrate cloud computing, sensors/IoT and semantic web aspects. In particular, one of the challenges concerned the transformation/adaptation of sensors data and metadata to a common semantic format. This process involved the design and implementation of middleware for the transformation of GSN virtual sensors data and metadata to semantic web metadata (compliant to W3C SSN). To this end, a sound understanding of semantic web technologies was required. The effective use of these technologies was associated with a significant learning curve for most of the participating researchers. Similarly, extensive use of semantic web technologies (notably of SPARQL) was made as part of the implementation of methods for dynamically accessing data and metadata in the cloud. Specifically, SPARQL queries had to be formulated and executed against the OpenIoT ontology (which was an enhanced version of W3C SSN ontology). In order to alleviate the complexity of learning and using SPARQL and other semantic web technologies, OpenIoT implemented middleware wrappers and visual tools, which hide the low-level details of the semantic web technologies.

Another implementation and deployment challenge concerned the deployment of several components of OpenIoT within cloud infrastructures. This involved the cloud-deployment of infrastructure elements and IoT data elements. The project experimented and successfully realized integration with public cloud infrastructures (like Amazon Elastic Compute Cloud), as well as with private cloud infrastructures (i.e. clouds built and operated by the project partners). OpenIoT gives therefore freedom of choice in the selection of cloud infrastructure.

Additional concerns were associated with the scalability of the OpenIoT system, as well as with the implementation of cloud-based concepts such as on-demand service formulation and «pay-as-you-go» operation. In order to address these challenges, OpenIoT introduced, specified and implemented a «Scheduler» component, which receives requests for cloud-based IoT services and accordingly discovers and reserves the resources needed to deliver the requested service. The Scheduler enables the handling of multiple concurrent requests to the IoT/cloud system, while at the same time making provisions for reserving and tracking the resources (e.g., sensors, devices) needed. The latter reservations form also a foundation for the implementation of utility-based mechanisms, since they keep track of the utilization of resources in the scope of a given IoT service.

The integration of the OpenIoT infrastructure has also been very challenging. To this end, OpenIoT has (early) on devised a novel architecture for IoT/cloud integration, which boosted modularity. OpenIoT has taken into account the concepts and principles articulated in the scope of the Architecture Reference Model (ARM) that has been introduced by the FP7 IOT-A project.

7.2.3 Project Results

The main result of the project has been the implementation of an open source middleware platform, which enables the development, deployment and operation of semantically interoperable IoT applications in the cloud. The architecture of the OpenIoT middleware platform is depicted in Figure 7.1:

More specifically, the architecture comprises three panes/layers, namely [11]:

- The physical plane, which deals with the acquisition of observations from the physical world, through either physical or virtual sensors. At

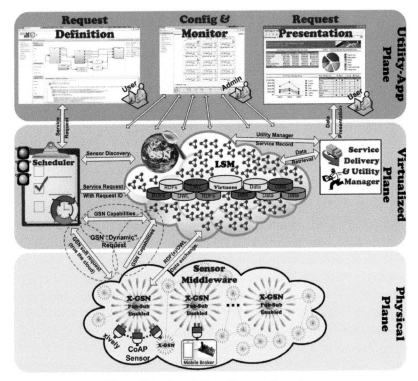

Figure 7.1   OpenIoT Architecture Overview

this level, an enhanced version of the GSN sensor middleware (called X-GSN) is used in order to integrate data streams from multiple sensors. The X-GSN platform undertakes to transform the data streams to RDF (Resource Description Format) compliant to the OpenIoT ontology, as well as to stream these data to the cloud infrastructure.

- The virtualized plane, which provides the means for discovering, accessing and processing IoT data in a semantically interoperable way. At this layer, data are virtualized, since they are represented in a common way regardless of their location and IoT source. The virtualized plane includes the scheduler component, which deals with the processing of requests for IoT services and the subsequent reservation of resources. It also includes the service delivery and utility manager component, which delivers IoT services according to the discovered sensors and calculates the relevant utilization of resources.

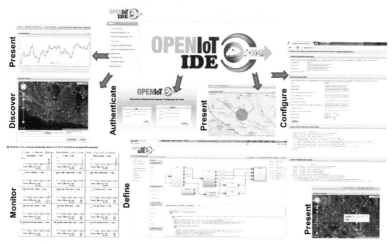

Figure 7.2 Overview of the Tools comprising the OpenIoT IDE

- The applications plane, which includes a range of development and configuration tools. The development tools («Request Definition») enable the visual (zero-programming) definition of IoT services, thereby enabling service development without deep knowledge of semantic web technologies (such as SPARQL). At the same time, the configuration tools («Config & Monitor») facilitate the monitoring of IoT data and services in the cloud. Note also that the applications plane includes also mechanisms and middleware for the visualization of IoT services («Request Presentation»).

The various development and management tools of the OpenIoT application plane are integrated within a single environment, which is conveniently called OpenIoT IDE (Integrated Development Environment). The OpenIoT IDE enables also the interactions of the various tools (see Figure 7.2), thereby delivering added-value over the stand-alone use of the tools.

OpenIoT is an open source project available at [32]. Some baseline statistics for the project (as of early February 2014) are listed in the table below.

Table 7.1 Information About the OpenIoT open source project

Number of Commits	Number of Contributors	Lines of Code	Estimated Cost (COCOMO model)[10]
960	13	177 621	28 man-years

OpenIoT is currently used for the development and integration of four proof-of-concept applications by the partners, in the areas of smart agriculture, smart manufacturing, ambient assisted living and mobile urban crowdsensing. Furthermore, it has been adopted as a baseline platform from the VITAL smart cities project [12]. Also, several independent researchers have downloaded and used OpenIoT for streaming and processing data from their sensors.

The project has recently received the BlackDuck open source rookie award, as being one of the top ten open source projects for 2013.

7.2.4 Acceptance and Sustainability

As an open source project, OpenIoT aspires to achieve wide adoption with the IoT open source community. Emphasis is therefore paid in community building activities, which attempt to stimulate the interest of open source contributors and IoT researchers. Special focus is paid on the networking and collaboration with other projects of the EU IERC cluster, given OpenIoT's strong presence in several working groups of this cluster. Following the development of a critical mass for the OpenIoT community, the sustainability strategy of the project involves the offering of services to individuals and organizations adopting and using OpenIoT. The services to be offered including training services, support services, as well as consulting services emphasizing on how to build and deploy IoT applications using the open source infrastructure of the project. The consortium is considering several ways for structuring and organizing these services, including the establishment of a new legal entity (either for profit entity, but possibly a non-profit foundation).

The partners have also actively pursued new collaborations (including participation in new EC funded research initiatives) towards continuing and sustaining the OpenIoT developments. While participation in such initiatives provides opportunities for enhancing the OpenIoT developments, the establishment of a dedicated entity for taking up OpenIoT is a more focused way for fine-tuning and perfecting existing developments and offerings.

7.2.5 Discussion

This section has presented the inception, evolution and main results associated with the OpenIoT project, which builds an open source middleware platform for building IoT applications in the cloud. OpenIoT has managed to develop and provide a first of a kind open source infrastructure, which guarantees the semantic interoperability of data streams and datasets that stem from diverse IoT systems. A key feature of this infrastructure is its ability to integrate these data sets within a cloud infrastructure, while also providing a rich set of tools for building IoT applications.

The novel character of the project, along with its open source nature, have already attracted the attention of several third-party stakeholders, which have already used it or planning to use it in the imminent future. Furthermore, the reception of the 2013 open source rookie award from BlackDuck has reinforced the momentum of the project through improving its brand and reputation, but also through raising awareness about OpenIoT's results world-wide. The OpenIoT partners are currently building on these early successes in order to maintain the momentum and evolve OpenIoT as a widely acceptable (de facto) open source infrastructure for building and deploying semantically interoperable IoT deployments. The consortium invites open source users and contributors to engage with the project, while it also pays significant attention in the reception of feedback. Such feedback is expected to be invaluable towards fine-tuning the project's developments and making OpenIoT even more appealing to the open source community.

7.3 iCORE

iCore project main focus is to enrich the Internet of Things with the use of cognitive technologies and enhance IoT-based applications to become more responsive and adaptable to changing user needs. In this section we describe the experience acquired along the full project execution cycle covering imple-mentation activities design, components and demo realisation, integration and validation of solutions, all the way through to transfer of results onto concrete, stakeholder supported trials.

7.3.1 Design

Since initial project setup and with a strong industrial representation in the consortium, iCore was designed to have a substantial part dedicated to generation of impact through proof of concept demos expected to cover a number of application domains, represent the interests of the involved parties and provide a suitable means to validate project results. The use cases selected to drive implementation in a demo style manner are briefly illustrated hereafter.

7.3.1.1 Smart home and assisted living

The Smart Home environment aims to leverage on IoT to improve the quality of life for the disabled and the elderly as well as provide monitoring tools for both family members and healthcare professionals. Additional objectives of this proof of concept include: (i) easy integration of care taking, (ii) remote health care, (iii) home automation and (iv) online medicine purchasing services. Additionally it creates possibilities of a new business eco-system among the elderly, the impaired, patients, family members, caretakers, doctors, nurses and pharmacy stakeholders.

7.3.1.2 Smart business and logistics

This use case describes how to best track goods that are transported from suppliers to retailer via a "mesh" of warehouses with road/air/sea transport in between. The real end-user issue is the lack of insight in the storage and transport conditions of goods between suppliers and consumer. To address these needs, a fine grained ICT monitoring system (e.g., a wireless sensor network based) is applied. E.g., a retailer wants to know if he can accept a shipment of temperature sensitive medicines, a transport operator wants to know if it can avert a claim of spoiled goods, since it thinks it kept the goods within the specified temperature tolerances and suppliers/retailers wants to know when to expect a delivery of goods. And in case of violation of storage and transportation conditions, parties responsible for the goods want to be able to act as soon as possible to reduce product spoilage (and associated claims).

7.3.1.3 Smart-city – transportation

This use case demonstrates the virtualization and use of ICT objects in the automotive industry, to create, configure and use mobility functions and services while driving and, in a seamless way, also in pre-trip and post-trip services. Major aspects and challenges are the availability of objects within the vehicle and from the outside world, considering the vehicle as a complex and autonomous eco-system and not an always-connected environment.

7.3.1.4 Smart meeting

The scope of the Smart Meeting use case is to provide meeting organizers with features and the capability of efficiently managing the whole meeting life-cycle from its organization, to its execution and the meeting wrap-up. As such the concepts and ideas pushed forward can be useful to a variety of interested parties ranging from small businesses and universities regularly hosting project meetings to large conference venues.

7.3.1.5 Rationale for chosen use cases

The reasons behind the choice of these use cases, which have driven implementation activities, have been threefold. On the one hand they had to ensure they would adequately highlight the concepts the project would propose solutions for. Also, to ensure enough momentum in their execution and wide impact and validity from a more industry oriented perspective, they had to reflect the interest of at least one iCore internal industrial Partner. This resulted in industry champions sustaining and driving each of these use cases. Finally, they had to also meet criteria for wider business appeal and showcase solutions to widely recognised problems, which would ensure resonance and interest beyond that of project participants and industry champions.

7.3.2 Project Execution

The biggest challenge faced during project execution related to implementation of selected use cases, even though at a small scale level such as can be the one which proof of concept demos are expected to cover, has been the integration amongst Partners from different organisations, located in different locations and having separate strategic goals.

More in detail at first contributors had to refine the scope to ensure continued commitment and ability to deliver according to own expertise whereas at a later stage the issues were rather associated with having to spend a substantial amount of effort to interfacing components contributed by individual Partners. Besides interfaces to be made available remotely across the Internet, also the protocols of interaction between components had to be harmonised to ensure proper integration of components. Heterogeneity of device imposed substantial integration effort having to install specialized, device specific software.

Besides integration issues, other problems have been faced such as having to protect privacy of user data, having to get approval for installing applications and services in tackled environments given a general mistrust in automation (i.e. having the man outside the loop in the service instantiation and setup).

As the project execution progressed and results became available for integration, a reassessment of accomplished and planned project implementation activities has been made. To maximise impact and involvement from various contributors towards a worthwhile and as widely as possible shared set of goals, further investment in terms of effort and budget reallocation was considered to strengthen the implementation activities, extending the collaboration outside the consortium and towards the creation of real-life trials involving concrete stakeholders and in some cases real end-users.

The decisions made half-way through the project execution supported the implementation of separate trials (as opposed to previously introduced use cases for demos) that complemented each other in a number of ways (explained in the remainder of this section after the description of the trials) and provided means for iCore results validation over a reasonably wide spectrum of areas, ranging from technical to business and more stakeholders oriented perspectives, as illustrated in the results section.

Extending the use of components from the proof of concept demos into more ambitious and extensive trials involved a great deal of interactions with interested stakeholders producing with time a clearer path for exploitation of project results. Besides this a number of other steps were taken to maximise impact, namely through a set of dissemination related activities including not only presence with iCore demonstration stands at international events and publications but also giving concrete opportunities to the IoT community of developers for re-using the achieved and showcased results. In particular some use cases results were made available as GNU General Public License code, as design and development guidance documents as distribution of well-defined Open APIs for developers communities to implement relevant solutions for experimentations.

7.3.3 Results Achieved

The main results of these implementation activities can be categorised in two main areas, which have led to showing results in small-sized proof of concept demos at first and within wider trials towards the end of the project.

The initial target for the first half of the project has been to showcase at conferences small demos. Such results have mostly recreated real-world situations and showed how an iCore supported system would react to support both, the needs of end-users as well as those of intermediate stakeholders playing a role in the value-chain associated with the improved IoT service. In particular these results have been acknowledged through awards achieved in two consecutive years at Future Network Mobile Summit for Runner-up Demo award and Best Demo award in 2012 and 2013 respectively.

The first demo showed how, through the concept of Virtual Objects (VOs), Real World Objects can be semantically enriched to foster their reuse and made to behave more autonomously i.e. generating events, notifications and streaming sensed data, which can be tailored to the needs of the applications that use them. The demo also presented how enriched objects can be combined dynamically and automatically to achieve more complex functionality to achieve better robustness of the IoT. These object "self-management" aspects were also used as the basis for the implementation of other use cases.

These results have been enriched with further features producing a more complete demonstrator that got the best demo award in 2013. In this case the above mentioned results were enhanced with the support of predictive models, able to reproduce Real World Knowledge and used in IoT applications that could adapt to the changing situation they were executed in. Those achievements have been at the core of one of the final trials proving amongst other things a more efficient usage of network and communication resources in the context of a smart-city surveillance application.

In the second part of the project execution, the iCore solutions have been transferred to more comprehensive and ambitious implementation activities involving real end-users and real stakeholders as illustrated hereafter in the description of the four iCore final trials, still on-going at the time of writing this section.

7.3.3.1 Smart tourism trial

The first trial is in the smart-tourism domain; it is located in Athens and involves a local travel agency as intermediary, as well as a user base of approximately 300 people from various tourist groups visiting different sites around the city. It exploits iCore solution in the implementation of three separate applications, one labelled "smart hotel", one "smart moving in the city" and one "smart tour in the city". In the first one users use IoT to control room conditions within given hotel policies and receive relevant notifications about their stay; the second application touches on smart transport issues and optimises a coach of tourists planned visits based on traffic and queuing at venues info; the third application uses IoT to send group/location based notifications to the users.

Envisaged collection of feedback from the users of this trial is expected to further help in the evaluation of iCore solutions used in it, providing valuable feedback for the "software industrialisation" of the iCore platform and for the improvement of iCore components and interfaces.

7.3.3.2 Smart urban security trial

The focus of this trial is on people safety (i.e., context of a VIP visits) and evacuation management in a smart urban area in case of threats such as toxic chemical cloud, crowds panic and aggressive people behaviour. Envisaged end-user roles are the VIP to protect, a dedicated VIP team that address the close protection, policemen within the area that contributes to enhance the VIP security and a police mobile Command and Control centre within a truck that coordinates the overall security stakeholders. The trial foresees the deployment of an IoT-based surveillance system which, leveraging on use of cognitive technologies for the enhancement and activation of relevant sensing capabilities, can be used to support decision making during crisis management.

This trial is promoted and supported by big industry players and is meant to illustrate how iCore predictive modelling can be used to support decision making and optimise the usage of network resources through situation-aware surveillance.

7.3.3.3 Smart asset management trial

This trial foresees the deployment of a "smart IoT" system able to continuously locate and assess status and maintenance needs of medical equipment in a large unit of a hospital and route operators to these in a situation-aware way. The trial aims to integrate iCore solutions with existing technology of a system integrator specialised in the real-time management of geospatial events. It validates how to exploit iCore composite virtual objects features to decouple pre-existing enterprise-level spatial data infrastructures from the actual objects generating events.

This trial is meant to show the value iCore can bring to SMEs in reducing time-to-market for deploying solutions for the management of spatio-temporal IoT generated events in a variety of application domains.

7.3.3.4 Smart amusement park trial

This trial is expected to disseminate iCore results also outside a European context, namely in China, showing international relevance of achieved results and in particular addressing the deployment of an entertainment application for theme park visitors. The envisaged applications are to support the production of multimedia souvenir to individuals as well as groups of tourists, leveraging on IoT technology and providing enriched experience including RFID-triggered video snapshots edited and provided at the end of the visit.

Similarly to the previous trial, this implementation activity is directly targeted at assessing the usability of the project results in an application context very close to a market product being promoted by an SME.

The achievement of such implementation results has also provided participants with many invaluable lessons learnt. Implementation, integration, testing and validation activities require a considerable amount of effort amongst the involved parties that should not be underestimated when projects proposals are prepared. Besides these considerations, planning for integration meetings and face-to-face interactions is a good practice for speedy solution to problems and for reaching goals within allocated time and budget. In this realm it is also important to proceed towards step by step integration of software modules, continuously and frequently testing for way forward implementation. Likewise, early exposure to stakeholders can ensure needed features are taken on board from the very beginning.

Another point to consider and thoroughly assess is the time at which implementation choices have to be made for the sake of minimising interoperability and integration issues amongst various partners' contributions. Based on the expertise available and on the interest of involved parties, the right balance must be pursued between too early choices on a single platform, which stifle exploratory alternatives and too late decisions on common interfaces which lead to many different low-impact demonstrators, defeating the purpose of deriving strong impact from collaboration.

7.3.4 Acceptance and Sustainability

As already mentioned the refinement of the use cases to be implemented and the transfer of results to more ambitious implementation trials included actions to engage the participants and ensure their commitment outside the context of the project, pursuing in this way wide acceptance of results. Worth mentioning in this context a set of specific activities (using the Value Proposition Canvas)

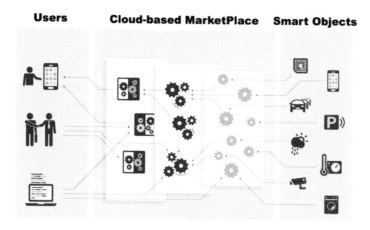

Figure 7.3 Conceptual level of COMPOSE MarketPlace

devoted to the assessment of the business value of iCore solutions in the context of the implemented trials and associated stakeholders, with results to be made available at the end of the project via a public deliverable. Preliminary results highlighted a prevalence of business value for iCore propositions which cannot be extracted by end-users (i.e., those installing apps on their smart devices) but rather have to be channelled through mid-tier providers who can enhance the features and the adaptability of the solutions they sell to those in the value-chain directly selling products and applications to end-users.

Besides extrapolating business value for targeted stakeholders and promoting acceptance (and hence sustainability) of iCore solutions this way, a certain degree of legacy after the end of the project will also be ensured through other alternative means. Looking at the associated activities and IoT strategies of various partners of the consortium, it is clear that the ones that will ensure adoption of results will be the SMEs involved in the project, for which the need to capitalise on the investments made throughout project duration is paramount. In some cases the deployed infrastructure installed for the trials will remain usable after the end of the project and reused and maintained in the context of further projects and experimentation.

To further support sustainability of project results most advanced results and components are to be released into vibrant IoT open-source communities, in the form of open APIs and GNU Licence software. The Smart Tourist application for example will be made available as an application in the Android marketplace and the most mature of the project results will also be leveraged upon by SMEs closely related to project partners.

7.4 COMPOSE

The vision of the COMPOSE project is to advance the state of the art by integrating the Internet of Things (IoT) with the Internet of Services (IoS) through an open marketplace, in which data from Internet-connected objects can be easily published, shared, and integrated into services and applications.

For developers, COMPOSE provides an open-source infrastructure and a set of tools and methods for building smart applications that can communicate with smart objects (smartphones, sensors, actuators) and external information resources. The key features of COMPOSE can be summarized into the following:

- Scalable, cloud-based infrastructure featuring Platform as a Service (PaaS) for hosting back-end applications and an IoT Marketplace.
- Provision of a set of tools (SDKs, IDE, recommendation engine, etc.) for developing smart applications that can communicate with external resources.
- Provision and integration of sensor communication technologies (Web-based bi-directional communication featuring advanced Web 2.0 technologies like Web Sockets).

7.4.1 Project Design and Implementation

The logical architecture of the COMPOSE platform is depicted in Figure 7.4. The main components of the framework are the COMPOSE Marketplace, the Run-Time engine and the Ingestion layer consisting of Smart Objects and services.

The COMPOSE marketplace implements a Service-Oriented Architecture, where any resource is provided and consumed in the form of a service. An Object is then elicited to a service object when it becomes accessible through a network connection. While an object would be the sensing device monitoring the status of a house, for example, its corresponding service object is the abstraction of a given feature provided, such as data on the temperature inside the house. Service objects will comply to the COMPOSE standardized interfaces, and will be potentially running the COMPOSE runtime environment in order to be (i) accessed from the Marketplace for gathering information (ii) actuated (iii) dynamically reprogrammed at run-time. Different interfaces will be defined in order to address objects heterogeneity. Service objects can be stand-alone or composite. Composite service objects are the aggregation or composition of simple ones. For example, the house service object is the aggregation of various objects providing information on temperature, presence, light, sound, and more. Composite service objects can provide information obtained from the aggregation of multiple data flows coming from different stand-alone service objects.

An object is any real-word active device capable of either providing contextual data or acting on the external environment. This includes sophisticated devices such as smartphones and multi-sensing platforms, but also simple ones like RFID tags and QR codes.

A service can be both a consumer of information originating from service objects and an actuator connected to one or multiple service object(s). When acting as a consumer, a service uses the information originating from one or more service objects to perform a given task. In contrast, when actuating on

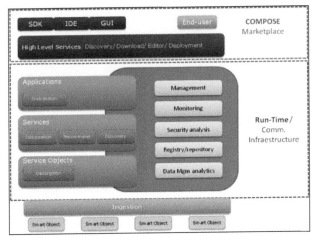

Figure 7.4   Main components of COMPOSE

a service object, a service issues a task to such a service object. For example, a service could first read (consume) the information from a light sensor in a house, and then determine whether to switch off (actuate) the light. Services can be simple or composite. Composite services incorporate the functionalities of other services and rely on them to properly function.

End users are the consumers of the services managed through the open marketplace. A user can be a person, accessing the marketplace through the installation of a given application on a personal device or computer, or a machine, through an appropriate machine-to-machine (M2M) protocol interface, interacting with the market to integrate IoT services into its business process.

The COMPOSE open marketplace is the distributed infrastructure orchestrating all aspects of the components mentioned above. Data coming from objects can be streamed into the marketplace, where their counterpart service objects will operate, and can be published such that other entities will be able to find the information and consume it. The marketplace will ensure that privacy and security aspects are well taken care of and additional non-functional requirements such as QoS may be specified.

A developer in order to build and deploy IoT applications on the COMPOSE architecture needs to: (i) Use the IDE and the SDKs and the high level services to discover existing service components or build new ones. The service components can be sets of classes or high level scripts that define what internal storage services will be used, how data will processed and stored within the infrastructures. (ii) Define and implement the communication with smart objects (like sensors and/or smartphones) and external resources. (iii) Deploy the services into COMPOSE infrastructure. The process is quite similar to deploying applications in cloud PaaS environments.

7.4.2 The IoT Communication Technologies

Interaction with smart objects and remote services requires the utilization of IoT technologies. Smart objects do not feature only sensing devices but can also integrate various actuators (switches, motors, relay circuits) that need to be communicated by the external services. Thus, bi-directional communication mechanisms are needed that take into account the limited resources of smart objects (low cost, low power hardware, etc.) and can also be deployed behind network firewalls, NATs, etc. For this purpose, COMPOSE is adopting the Web of Things notion [13]. Each smart object is considered as a web-enabled object that can communicate over HTTP and consume REST web services. For bi-directionality, Web Sockets offer the ability to back-end services to send notification to clients (i.e., connected web objects) when needed. Clients do not need to continuously poll the servers for updates, neither to be reachable by Web (i.e., open to HTTP connections) that could be in many cases (private networks or networks over 3G) not feasible. In addition, popular binary protocols, like MQTT will be also integrated to allow remote interaction with devices that have low resources (e.g., battery powered actuators).

7.4.3 Execution and Implementation Issues

7.4.3.1 The COMPOSE services

The COMPOSE Services are the software components that allow the execution of back-end applications on the COMPOSE infrastructure. The infrastructure is hosted on a Cloud-based scalable environment based on Openstack [14] and Cloudfoundry [15]. Developers can port their own applications to the proposed infrastructure by using a number of different programming languages (Java, PHP, Ruby, Node.js, etc.) and libraries for data storage and communication.

7.4.3.2 The back-end technologies

The framework is mainly built on top of existing state-of-the-art technologies. In particular, in the scope of the Objects as a Service work, the following technologies are being integrated:

REST (Representational State Transfer) is the architectural principle that lies at the heart of the Web, and uses HTTP to provide application level transport.

CouchBase Server [16] integrates an in-memory key/value store and a NoSQL back-end (CouchDB [17]) to provide a novel approach to the field of horizontally scalable databases.

The two dominating distributed stream processing frameworks, Apache S4 [18] and BlackType Storm [19] are key technologies in the framework. They are leveraged to perform the automatic translation between service-defined data management primitives into stream processing graphs.

Currently different options to define DSLs are being explored, from more static and simple solutions such as Apache Pig [20], to more complex approaches such as the Scala framework [21], which provides language virtualization.

7.4.4 Expected Project results

COMPOSE aims to deliver a highly scalable sensor information streaming and processing platform, interfaces for creating and deploying smart applications and services and tools for service annotation and discovery.

The project is also realized through three different pilots, namely the Smart Spaces, Smart City and Smart Territory that aim to demonstrate the features of the platform.

7.4.4.1 Smart Spaces

The Smart Spaces scenario focuses on IoT-based services for indoor environments such as, e.g., retailer stores, office or home environments. Specifically, the pilot addresses the Smart Retail application scenario with two specific objectives: a) The augmentation of the user shopping experience through enhanced interaction with displayed products and personalization of in-store delivered services. This will support the delivery of advanced in-store services, which eventually will improve the quality of the in-store shopping experience of customers. b) The development of an in-store analytics platform in order to precisely model the in-store behaviour of customers. This will serve as the basis for taking informed decisions on the store management, thus increasing the quality of the delivered service and eventually the profitability of the store.

7.4.4.2 Smart City

The Smart City pilot takes advantage of the existing Abertis SmartZone and the OpenData project of Barcelona and will use the COMPOSE Open Marketplace for developing and delivering new services for its citizens. Special focus will be given to new opportunities derived from cross-data and its integration in the day-by-day life of the people living and working in the city. The creation of these new opportunities will take advantage of existing infrastructures and information. Available data on public transport and the electrical vehicles use will also be considered in the scenario. The pilot may involve several of the following use-cases: a) Car sharing and multimodal route planner - Using an Electric Vehicle car sharing service, checking the traffic status and forecast to decide the best route and the available parking spaces in the destination. Also, users can choose between public transportation, a shared car, or taking the private car. b) Weather and pollution monitoring – choose the best jogging route at a particular time.

7.4.4.3 Smart territory

This pilot focuses on the implementation of the identified use cases related to the Tourism sector in Trentino. The COMPOSE marketplace will be piloted to be an entry point for smart and personalized service mashups allowing tourists to explore the sport and cultural touristic facilities of Trentino and leveraging various available infrastructures in Trentino: meteorological sensors, touristic resort sensors, cultural events, people carrying smartphones, etc. Among the stakeholders the pilot will involve Riva del Garda Tourism Board, Trentino Network communications infrastructure provider, Meteotrentino climatic data provider, and users carrying smartphones.

7.5 SmartSantander

SmartSantander project has created an experimental test facility for the research and experimentation of architectures, key enabling technologies, services and applications for the Internet of Things (IoT) in an urban landscape. The facility has been conceived as an essential tool for achieving the European leadership in Internet-of-Things technologies, which would permit the scientific community to experiment and evaluate services and applications for smart cities under real-life conditions. The project committed to the ambitious deployment of 12 000 IoT devices in Santander, a small-medium city of 179 000 inhabitants located in the North of Spain. Apart from the devices installed in Santander, the project has deployed devices in cities of Belgrade, Guildford and Lübeck, permitting the federation among different IoT test-beds around Europe.

7.5.1 How SmartSantander Facility has Become a Reality?

SmartSantander facility provides a twofold exploitation opportunity [22]. On one hand, the research community gets benefit from the massive deployment of IoT technology in such a unique facility, which allows true field experiments in the real world. On the other hand, different services fitting citizens' requirements have been implemented and validated under real conditions in the urban landscape.

The facility encompasses IoT technologies, in different areas of the city, with various applications domains ranging from public transport, urban services such as waste management, parks and gardens irrigation, public places and buildings, work and residential areas, thus creating the basis for development of Santander as a smart city. The areas of deployment have been selected based on their high potential impact on the citizens, driven by the city of Santander requirements and strategy, thus validating the acceptance of IoT based services in real life environments, exhibiting also diversity, dynamics and scalability.

The deployments and further developments were organized in three phases:

- **Phase 1:** 2 000 IoT devices, including repeaters and Gateways (GWs) that allows creating mesh networks, providing basic experimentation support together with outdoor parking and environmental monitoring services.
- **Phase 2:** 5 000 IoT devices, adding more heterogeneity to the facilities and providing advanced tools for the experimentation. Furthermore, additional services were implemented: environmental monitoring with fixed and mobile nodes, traffic monitoring, and guidance to parking lots, parks and gardens irrigation, augmented reality and participatory sensing.
- **Phase 3:** 20 000 IoT devices, supporting the federation of nodes deployed in the different cities as well as those with other Future Internet and Research Experimentation (FIRE) facilities. Regarding to the research and experimentation issues, advanced cross-testbed tools were created. Services developed in the previous phases were improved with new deployments and technologies.

Nowadays, not only the initial goals have been achieved, but also the city of Santander has been placed itself in the forefront of technological innovation, thanks to the SmartSantander initiative.

7.5.2 Massive Experimentation Facility: A Fire Perspective

Aligned with the FIRE initiative, the facility offers to the research community the possibility of experimenting on top of the deployed nodes, in two main ways [23]:

- **Native experimentation.** Most of IoT bodes, those with fewer constraints in terms of battery, can be flashed through Over-The-Air Programming (OTAP) or Multihop OTAP (MOTAP), as many times as required and with as many different experiments as it might be needed. In this sense, researchers can test their own experiments, such as routing protocols, data mining techniques or network coding schemes.
- **Experimentation at service level.** Data generated by IoT devices is also offered to the researchers, letting them to combine and correlate information aiming at conceiving smarter and more sustainable services within the urban environments.

Besides the synergies with the traditional experimentation frameworks mainly coming from the FIRE community, it has been also identified a good collaboration path with the FI-Lab promoted by FI-WARE [24]. Nowadays, businesses and entrepreneurs from all over the world can develop services taking advantage of the infrastructures deployed in SmartSantander.

7.5.3 City Services Implementation: The Smart City Paradigm

The smart city paradigm spans across many subjects both technological and sociological. Citizens play a major role in this paradigm as the final recipients of the services supported by the associated infrastructure but also as key drivers for continuous innovation. In this sense, in order to meet tangible requirements it is important to involve them so as to consider their personal opinion when ranking different kinds of services. Following this approach, the SmartSantander project analysed, designed and developed the services that were interpreted as a priority by the local authorities, regional government and end users.

7.5.3.1 Parking service management

One of the most common use cases when considering connected cities addresses integral traffic control. Among others, management of limited parking including specific spaces reserved for people with disabilities, control of load and unload areas and traffic prediction are the most relevant scenarios to be considered.

While many cities worldwide provide information about indoor parking lots, similar information for outdoor parking areas is rarely available. With the aim of reducing CO emissions and other pollutants, as well as petrol consumption, IoT technology, characterized by its pervasiveness, has become a very attractive solution both technically and economically speaking. The deployment of IoT technology, making publicly available the information about how many parking spots are available in a specific area and how to reach them, allows the drivers to reach available spaces in a much more efficient way. Moreover, it improves the exploitation of the parking service as it is possible to create occupancy models which are useful for further studies in terms of traffic prediction.

Around 650 ferromagnetic parking sensors have been buried under the asphalt of parking areas at Santander downtown streets (so-called Santander Zone 30) in order to detect the occupancy degree of determined parking lots. Furthermore, 10 panels have been deployed at the main roads and intersections of this area in order to guide the driver to available parking places within the different streets in the area. Both guidance panels as well as sensors installation process are shown in Figure 7.5.

7.5.3.2 Traffic intensity monitoring

Nowadays, the measure and classification of vehicles in road traffic is accomplished by inductive loops placed under the pavement. These inductive loops allow monitoring vehicle passing by means of different configurations,

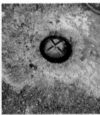

Figure 7.5 Examples of parking sensors installation and guidance panels

which provide with some of data in order to control several parameters of the traffic (vehicle speed, traffic congestion and traffic accidents, among others).

However, these systems have several constraints such as their deployment and, maintenance costs as well as to put into gear. Within the project, a solution based on wireless sensors has been deployed, creating traffic services such as a traffic status map that combines the information collected by the wireless sensors deployed in different road lanes (car speed, occupancy and vehicles count) with the information gathered by the legacy systems.

7.5.3.3 Environmental monitoring

More than 1 000 fixed nodes have been installed in street lamps and wall facades within the city of Santander, which monitor CO index, temperature, noise level and light intensity. Located at downtown, it is very representative and has been extended to other areas of the city by using devices installed in vehicles.

The new hardware deployed on public vehicles (buses, taxis, etc.) collects geo-positioned measurements of environmental parameters such as NO_2, CO, O_3, humidity, temperature, among others every minute. Thus, mobile nodes contribute to extend the facility to other parts of the city, covering a much wider area on a much more efficient way. Additionally, devices installed on mobile nodes may interact with the nodes placed at street lamps and facades, allowing researchers to carry out experiments on mobility.

Figure 7.6 shows some examples of the sensors installation in street lamps, facades, buses as well as park and gardens vehicles.

7.5.3.4 Parks and gardens irrigation

Traditionally, irrigation systems are managed in a quite static manner, without considering real-time parameters from each of the areas of the parks where usually exist different types of vegetation. However, within the project, IoT

Figure 7.6   Environmental monitoring fixed and mobile nodes

technology has been deployed to complement the automated irrigation systems currently used inside the three major parks of Santander: the Las Llamas Park, La Marga Park and Finca Altamira, covering an area of 55 000 m^2.

The implemented solution provides relevant real-time information to the garden management responsible and parks' technicians, aiming at improving performance and reducing exploitation costs of the park. IoT devices with special agricultural sensors measure parameters such air temperature and humidity, soil temperature and moisture, atmospheric pressure, solar radiation, wind speed/direction, rainfall all the nodes transmit wirelessly the data acquired to the SmartSantander platform.

7.5.3.5 Citizens apps

A great effort has been dedicated to provide citizens with applications that improve and ease their live in the city. Two applications can be highlighted.

SmartSantanderRA [25] is a free App, available both for Android and iOS platforms that uses Augmented Reality technology to present information about the city in a context-sensitive, location-aware manner to citizens and visitors.

Aiming at improving the user experience when visiting the city, the App unifies different data sources already available in the city, allowing end users to quickly and homogeneously access to information such as cultural agenda, shopping, transportation, touristic, public transport, tourism activities. Furthermore, the service has been improved with the deployment of 2 600 stickers with dual tags (NFC and QR codes). Figure 7.7 shows the installation of tags along the city as well as the promotion of the AR application carried out by the municipality. Around 415 tags have been placed in bus stops, allowing the users to know in real time how much time they have to wait for next bus (all this without the need to install a panel which is much more expensive). Around 2 000 tags have been installed in shops, aiming at stimulating and

Figure 7.7   Tags installation and application promotion

making this sector much more dynamic. Shop owners can now update the information on a daily basis with additional details of the shop (e.g., bargains, offers, exclusivities).

In around 18 months, more than 19 000 users have downloaded the application, having more than 810 000 accesses to the information provided by both the App and the tags.

Pace of the city [26] is an App based on the participatory sensing paradigm that has a dual profile. Firstly, citizens, through their smartphones, can report events and incidences occurring in the city (i.e., hole in the pavement). Secondly, it allows them to send observations from the sensors embedded in their smartphones periodically to the SmartSantander platform. Figure 7.8 shows both sensor measurements and incidences reported by the citizens within the city of Santander.

The Pace of the City service interacts with the municipality systems so that events reported by the citizens are automatically transferred to them. At the municipality, the incidences are processed and assigned to the corresponding team that deals with their solution. This information is also made available to the citizens, which can trace how the incidences are solved in real time.

The service has obliged the city council to completely change the way in which the city deals with the incidents reported by citizens. Now, most of the incidences are solved in less than 6–7 days whilst in the past it took them around four weeks.

Furthermore, the most popular regional newspaper also interoperates with the service allowing journalists to include geo-located pieces of news. Thus, in case of an accident, if a journalist is in the vicinity, he/she is able to write about it. All the users subscribed to these notifications will receive such information.

Figure 7.8   Participatory Sensing events and measurements

Currently, more than 7 000 users have downloaded the application and are actively using it. Finally, the analysis of the data collected by citizens behaving as human sensors provides a sign of how the city is evolving.

7.5.4 Sustainability Plan

The SmartSantander sustainability model has been conceived based on the creation of an urban smart city platform in which all the urban services must be integrated, as they are offered in public procurements. It combines the exploitation of the platform, mainly by private companies or entities, which would try to obtain commercial benefit from the exploitation, while maintaining ownership of the infrastructure within public agents.

The sustainability model follows a "holistic" perspective. It brings together actors from the public and private sectors and covering both the commercial and scientific scopes of the initiative. The model has been built based on the role carried out by four different parties:

- Public agents representing the ownership of the platform (which is cur-rently shared by Santander municipality and University of Cantabria), the regulatory agents, the citizens' representatives, and the pure research and scientific interest. In this context, the commitment from the municipality to incorporate IoT technologies in the next public procurements for the exploitation of the urban services results of special interest and emerges as a key mechanism for the viability of the SmartSantander facility.
- The urban service provider is another key component of the strategy, based on future calls for tenders that will release the municipality. These agents, acting as operators or concessionaires of various urban services, will contribute to the sustainability of the platform in a seamless and fair manner (as the new services rely or take benefit from the facility).
- The network operator and technology provider will also contribute to the economic sustainability of the platform, but always maintaining a win-win paradigm in which the operator may obtain profit from the registration of devices, certification of technologies, integration/use of the platform, and M2M-related incomes (SIM cards, traffic rates, etc.), among others.
- The experimenter is not envisioned only as a "monetary source", but also as a knowledge contributor. The incorporation and inclusion of

experimenters will enrich the platform with the latest technologies and will keep it updated in the scientific arena.

Finally, another relevant aspect, that must not be forgotten when assessing the sustainability of a project such as SmartSantander, is the fact that it must be perceived as an "added value" service by the different stakeholders (public or private agents; not end users) involved in the development, deployment and funding of the service. If commercial profitability (and not mere "sustainability") is mandatory for all the involved players, the chances of success will be more limited. As a consequence of the previous point, all other "non-monetary" benefits, such as the environmental or lifestyle improvements for citizens are also important aspects in the decision to release and maintain this kind of systems and services.

It is not only a matter of introducing cutting-edge technologies that contribute to have an increasingly modern city, the aim is to transform the cities into more liveable spaces in which "smart" and "non smart" citizens live. This constitutes an integrating view of the city, in which the society needs to be trained to make use of the technologies and infrastructures that the city makes available.

7.6 FITMAN

7.6.1 The "IoT for Manufacturing" Trials in FITMAN

The mission of the FITMAN (Future Internet Technologies for MANufacturing industries, [27]) project is to provide the Future Internet Public Private Partnership with 10 industry-led use case trials in the domains of Smart, Digital and Virtual Factories of the Future.

The FI PPP Technology Foundation project FI-WARE [24] identified in its Chapter III named "Internet of Things (IoT) Services Enablement Architecture" some basic components (called Generic Enablers), in order for things to become citizens of the Internet – available, searchable, accessible, and usable – and for FI services to create value from real-world interaction enabled by the ubiquity of heterogeneous and resource-constrained devices. On the top of these basic building blocks, FITMAN is developing some IoT-enabled Specific Enablers, customised for the Manufacturing domain (e.g., the "Shopfloor Data Collection", the "Secure Event Management" or the "Data Interoperability Platform and Services") and to be experimented in 10 industrial trials spread over Europe.

FITMAN Trials (4 conducted by Large Enterprises, 6 by SMEs) are expected to test and assess the suitability, openness and flexibility of FI-WARE Generic Enablers while contributing to the STEEP (social-technological-economical-environmental-political) sustainability of EU Manufacturing Industries. The use case trials, classified in Smart (shopfloor automation and control), Digital (product life cycle management) and Virtual (supply chain and business ecosystems), belong to several manufacturing sectors such as automotive, aeronautics, white goods, furniture, textile/clothing, LED lighting, plastic, construction, and manufacturing assets management.

The relevance of IoT technologies depends on the trial and in this section we are going to describe two of them, one in the automotive and one in the white goods manufacturing sectors.

7.6.2 FITMAN Trials' Requirements to "IoT for Manufacturing"

Future Internet technologies (BigData, Cloud Computing, Mobile web Apps, etc.) offer manufacturing industries the possibility to engage in a digital transformation leveraging advanced business processes [28] . Nevertheless, the objective of the FI technologies is not to replace the existing platforms and products, but to enable next generation business processes adding and implementing innovative aspects.

In order to reach these digital transformation challenges, the FITMAN technological architecture have been defined; i.e., the high level features leveraged by the FITMAN reference platforms and associated GEs/SEs to effectively support next generation business processes.

- In particular, smart trials are characterised by support of cyber-physical systems, overcome data discontinuity, facilitate production capacity as a service, make ramp-up production activities and operational routines much easier and more efficient, support self-tuning, self-diagnosing and optimizing features of modern process control, support advanced human-computer interaction, support human-centric ergonomic manufacturing process implementation, and secure data handling.

- On the other hand, the digital trials demand highly modular event-driven architecture, interoperability with major PLM platforms, quick, flexible, managed and intelligent integration of information, real-time linking with shop floor (in-field) data, secure (trusted) data management apps, quick development of customised user-focused (engineer, customer, production manager...) mobile and collaborative decision support apps, effortless development of advanced data analytics and (mobile) data visualization and cost-effective service operation and maintenance.

FITMAN "IoT for Manufacturing" trials support the transition towards self-organising production capabilities that leverage far more adaptive advanced manufacturing production sites.

7.6.3 The TRW and Whirlpool Smart Factory Trial

On one hand, TRW trial aims to develop a new generation of worker-centric safety management system in order to reduce the accident and incidents in the production workplace. The traditional prevention strategies are not capable of customizing specific plans and current human-based surveillance is not completely effective. Thus, the trial specially demands the functionalities of the real-time detection of ergonomic risks, as well as continuous data processing for events creation and corrective actions performance.

On the other hand, WHIRLPOOL trial is taking place in the context of the washing unit production line in Naples. The present scenario is characterized by an underutilization of the data gathered in the production line and unexploited benefit in terms of speed of reaction and effectiveness of decision taken by the factory staff. The overall objective is to allow the factory to reduce defects due to operation going out of control by preventing as much as possible machine interruption. This can be achieved enabling the decision maker with a mobile device and a system able to gather all the basic events happening along the production line, filtering them applying a selective algorithm and deliver enriched information about the event through the mobile device.

The TRW Trial Platform (see Figure 7.9) exploits FI technologies mainly related to IoT services. Going into deep detail, IoT Gateway Data Handling GE detects ergonomic risks in the shopfloor in order to avoid problems in the efficiency of the network. Additionally, the IoT Backend IoT Broker GE and IoT Backend ConfMan Orion Context Broker GE support end-user (e.g., prevention technician) to register, query, subscribe and update the data/events received.

WHIRLPOOL Trial is focused on the acquisition, real-time processing and dispatching of events originating from IoT sensors. In brief, the Smart Factory Platform instance (see Figure 7.9) deployed in the Trial is composed by IoT Gateway Data Handling GE that executes the custom event processing logic, evaluates conditions based on event payload, and generates new events when conditions are met; IoT Backend ConfMan GE that coordinates multiple event

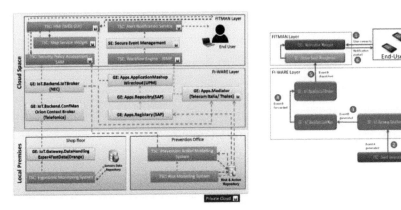

Figure 7.9 FITMAN IoT enabled Smart Platform at TRW (left) and Whirlpool (right)

producers on the shopfloor; and IoT Backend IoT Broker GE that coordinates multiple event consumers.

Additionally, both trials use Secure Event Management SE that enable the flexible and secure management of event distribution lists.

These trials support the introduction and acceptance of IoT technologies in the manufacturing industry, providing the necessary balance between security concerns and privacy concerns. More concretely, in TRW the implementation of IoT monitoring technologies supports the innovative human-in-the-loop model, getting a participative approach for the workers' empowerment solution.

7.6.4 FITMAN Trials' Exploitation Plans & Business Opportunities

The IoT technologies are expected to be the next revolution following the World Wide Web, providing new bridges between real life and the virtual world. This is an inflection point where technology is changing how manufacturing is being done, is being driven by the convergence of integrated control and information technologies. That in turn is being propelled forward with the IoT [29].

However, several key technologies must still be developed enabling the IoT vision to become a reality. FI-WARE GEs and FITMAN SEs are trying to address these opportunities, taking advantage of this considerable potential to improve productivity in the production process, from logistics and supply chain management to production process automation in the shop floor.

Regarding data collection, in the FITMAN Smart trials a huge amount of data are generated by sensors and intelligent objects in the shopfloor and they need to be automatically managed and in strict real time constraints. Digital trials require seamless and efficient interoperability between Real World, Digital World and Virtual World data and processes.

Furthermore, huge amount of data would be generated by the IoT devices, which would be needed in collaborative decision making situations, supporting predictive analysis and human decision making. FITMAN Smart trials pursue the improvement of product quality through a better and faster detection and more effective resolution of problems in the production environment, as well as a better integration of workforce in decision phases of the production process.

Finally, there is an opportunity to provide an integration backbone platform supporting different communication standards and protocols, and allowing different architectures to communicate with other networks. FITMAN Virtual trials are focused on the uniform management of very diverse and heterogeneous resources coming from different industrial sectors and application domains, without compromising safety and privacy requirements.

As a conclusion, IoT will enhance the way manufacturing companies work by saving time and resources and opening new innovative and competitive opportunities. FITMAN trials demonstrate how FI IoT solutions will be exploited by manufacturing industry with the objectives of improving internal processes, reducing costs, increasing safety & security, increasing efficiency, reducing time to market, and improving quality, among others.

7.6.5 Conclusions and Future Outlook

The FITMAN project is a technology trialling initiative, where research results (in this case also developed by other projects in the FI PPP) are tested and experimented in realistic scenarios. Lessons learned, best practices and identification of new business opportunities will follow the experimentation towards the end of the project (beginning of 2015), so it is premature now to anticipate considerations about IoT impact to the manufacturing industry. What we can say is that in FITMAN we just focussed on IoT adoption in Smart Factory environments, as testified by TRW and WHIRLPOOL trials, but in the Factories of the Future PPP some other projects are adopting IoT in Digital Factory contexts: product / assets tracking & tracing along the value chain, predictive diagnosis and maintenance, training, recycling and re-manufacturing. In our opinion the whole domain of "IoT for Manufacturing

Enterprises" is going to take momentum not just in the research & innovation communities, but mostly in enabling new business models and business opportunities for EU manufacturing industry and its IoT solution providers.

7.7 OSMOSE

The main objective of the OSMOSE (OSMOsis applications for the Sensing Enterprise [30]) is to develop a reference architecture, a middleware and some prototypal applications for the Sensing-Liquid Enterprise, by interconnecting Real, Digital and Virtual Worlds in the same way a semi-permeable membrane permits the flow of liquid particles through itself.

The following metaphor can be used to explain the concept: Let us imagine the Sensing-Liquid Enterprise as a pot internally subdivided into three sectors by means of three membranes delimiting the Real-Digital-Virtual sectors. A blue liquid is poured into the first sector (Real World - RW), a red liquid into the second sector (Digital World - DW) and a green liquid into the third sector (Virtual World - VW). If the membranes are semi-permeable, then following the rules of osmosis the liquid particles could pass through them and influence the neighbouring world, so that in reality the blue RW also would have a red-green shadow ambassador of the DW/VW, and similarly for the other Worlds. An entity (person, sensor network or intelligent object) in the blue RW should have control of their shadow images in the red DW and in the green VW, keeping them consistent and passing them just the needed information under pre-defined but flexible privacy and security policies.

The Sensing Enterprise will emerge with the evolution of IoT, when objects, equipment, and technological infrastructures exhibit advanced networking and processing capabilities, actively cooperating to form a sort of 'nervous system' within the next generation enterprise. The Liquid Enterprise is an enterprise having fuzzy boundaries, in terms of human resources, markets, products and processes.

7.7.1 The AW and EPC "IoT for Manufacturing" Test Cases

A world-wide helicopter manufacturer and its training simulators' ecosystems, as well as a global automotive camshafts manufacturer and its quality assessment value network will validate both concepts and measure business benefits and performance indicators.

AgustaWestland (AW), a world leader in rotary wing type and role conversion training, with over 50 years' experience of delivering high quality training to aircrew and technicians, is proposing the aeronautics test case. The flight simulators represented in Figure 7.10 are very important for pilot training, leading to higher level of training when compared to the real helicopter. They enable to reproduce several flight scenarios and conditions gathered during flight tests. However, to avoid negative training, especially for beginner pilots, it is extremely important to maintain the reliability, in terms of closeness to reality. It is needed to identify discrepancies (snags) from what is the expected behaviour comparing with the real Helicopter. In terms of maintenance, simulators are used by end-customers with a contractual commitment that requires an extremely high availability (close to 24/7), which can be provided only if all the logistic chain and support personnel are "synchronised" to the tasks that have to be performed.

By means of the OSMOSE project the reliability and availability of simulators will be addressed with the objective to improve the processes of information exchange and reaction among the different worlds. During the simulation the events providing data coming from the RW (pilots, simulator, sensors, etc.), VW (simulation software, virtualized sensors) and DW will be collected in order to improve the knowledge about the simulation and the simulation experience. Data collected will start and feed the processes of analysis and management of the simulator in relation with the availability

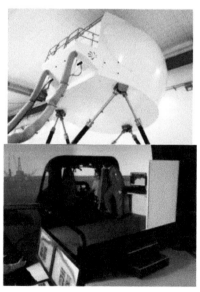

Figure 7.10 AW109 LUH Flight Simulation Training Device (full motion and fix based)

(e.g.: fault prediction) and reliability (snag assessment). Processes can include automatic, semi-automatic and manual steps depending of the initiating events and lead to improvement of the simulator availability and reliability, for example in terms of predictive hardware maintenance, software fix, etc.

Engine Power Components, GE, S.L. (EPC) is one of biggest equipment vendor of engine camshafts worldwide. Apart from providing camshafts for the transport sector, EPC is specialized in the market niche of big camshafts, more than 2 metres long, addressing applications such as naval industry, agricultural, military, generators, etc., developed for the main motor manufacturers (Cummins, Mitsubishi, John Deere, Caterpillar). They are proposing the automotive test case. Following the traditional manufacturing process of a camshaft (top of Figure 7.11), if any problem arises with the camshaft once it has been sold and installed, e.g., in an oil station far away, it will be necessary to travel there and physically inspect what is going wrong, stopping the production and causing important losses for both EPC and the customer. Thus the focal point of the test case is on the traceability over the set of operations involved in the camshaft manufacturing. OSMOSE will interconnect Real, Digital and Virtual

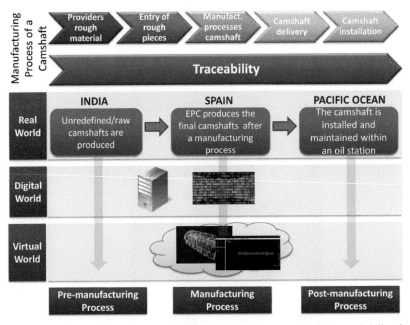

Figure 7.11   Traceability over the whole business flow of manufacturing and delivering a camshaft: correlation among the three worlds

worlds in order to keep this traceability, by introducing the innovative concept of the «DNA» of the part, linking features of the raw material, manufacturing and measurement information to each camshaft. EPC will be able to access a holistic signature of the product life-cycle within the liquid enterprise. By means of OSMOSE, it will be possible to manage the whole camshaft production process, being informed about the status and place of the raw piece, detecting possible failures of the manufacturing machines, maintaining a digitalized copy of the piece for quality control, becoming aware of exactly which camshaft is delivered where, as well as analysing post-manufacturing problems without the need for physical travels.

Summarizing, industrial validation and adoption of the two basic OSMOSE concepts in the two use cases implies the following application challenges: The manufacturer's knowledge will liquidly flow along the value chain and the lifecycle of the product; sensing the remote product will allow a more accurate decision-making process regarding diagnosis of anomalies and maintenance interventions; RW entities such as helicopters, pilots, camshaft CMMs, will be smart-connected to their DW representations in IT models, databases, multimedia repositories and their VW projections in what-if scenarios, simulations, forecasts.

7.7.2 OSMOSE Use Cases' Requirements to "IoT for Manufacturing"

Taking as a reference Use Cases' industrial requirements [31], OSMOSE needs to develop Big Data Analysis technologies, Predictive Maintenance, Assets Information Management and Global Assets Monitoring applications, all strongly supported with Interoperability, Trust and Security approaches for deployment into manufacturing enterprise systems. This makes OSMOSE a catalyst for the IoT implementation. These technologies and approaches handle interactions between the real/physical, the digital and the virtual assets of an enterprise that are considered the 3 worlds of the FI networked society. These assets empower various data and information sources (e.g., videos, web data, sensors, simulation) consumption, which intends to act similarly as senses do with humans, but in relation to an enterprise system, potentiating its knowledge acquisition and understanding to enable a context awareness status able to accomplish efficient decision making for further actuations.

With IoT for manufacturing, OSMOSE increases the enterprise ìs market value, by providing a mean of predicting behavioural dynamics of the domain of interest and actuate accordingly. This is achieved by aggregating information and knowledge from sensors, databases and contextual knowledge bases, analyzing it and by using the output to support and optimize the decision-making process. Proof-of-concepts are designed following a workflow of events and service calls that can be iterating among the 3 different worlds, and accessible at certain points of time through the Liquid Stargate. Event-processing technologies are natural to deal with data proceeding from real devices or from digital/virtual interfaces. These data streams are received into event-processing engines, in which pattern detection is applied and events are emitted to external actuators or systems. Controllers autonomously actuate over other devices to execute specific tasks accordingly to specific context awareness. Also predictions can be accomplished through appropriate simulation systems where smart objects, interact and negotiate. The Liquid Stargate is an architectural component that should enable enterprises to cross dimensions, reconfigure its systems, and provide access to information through an integrated multi-world view.

7.7.3 OSMOSE Use Cases' Exploitation Plans & Business Opportunities

The project plans to evaluate the test case trials impact in business terms, on the basis of the following benefits and its business indicators: Enterprise flexibility (measured through time and costs required to set up a collaboration form); Reduction of barriers to enterprise collaboration (indicators based on Maturity Models); Ability to exploit new business opportunities (number of additional business opportunities analyzed and implemented); Reduction of the cost of interoperability (cost of service composition and of data mediation and reconciliation); Advantages in time to market for new innovations (typical indicators are in the improvement of human competencies and in the value of innovations and RTD investments); Access to new markets (common standards); Access to technologies, knowledge, skills and information; Quality of FI solutions in order to make ICT product/service providers more competitive on the global market; Performance measurement; Reduction of barriers to geographically distributed team work; Access to innovation ecosystems, particularly for SMEs.

Accurate exploitation plan will ensure that the results of OSMOSE will find their route towards the market. A variety of exploitation modalities and channels will be adopted to ensure maximum impact. The OSMOSE Exploitation Strategy will follow the path described in Figure 7.12, showing

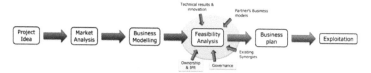

Figure 7.12 OSMOSE Exploitation Strategy

that besides the usual activities of market & business models analysis, the core of work will be delivered in the feasibility analysis where governance models and ownerships models from the OSMOSE partners will be scrutinized.

7.7.4 Conclusions and Future Outlook

The current stage of the OSMOSE project (started in October 2013) describes in detail the application expectations and challenges and the requirements for the main research results: the osmosis Middleware, the liquid Real-Digital-Virtual Stargate and the two sensing applications in aeronautics and automotive. Future business opportunities still need to be identified, but preliminary investigations show us that the osmosis approach could be a valid solution to achieve world's interoperability also in other domains, such as Smart Cities or Smart Home applications.

Acknowledgments

Many colleagues have assisted and given input to this chapter. Their contributions are gratefully acknowledged.

List of Contributors

Alicia Gonzalez, ES, Innovalia Association, OSMOSE
Antonio Collado, ES, Innovalia Association, OSMOSE
Carlos Agostinho, PT, UNINOVA, OSMOSE
Charalampos Doukas, IT, CREATE-NET, COMPOSE
Claudio Pastrone, IT, ISMB, ALMANAC
Dan Puiu, RO, Siemens, iCore
Dimitris Kelaidonis, GR, UPRC, iCore
Eva Vega, ES, ATOS, FITMAN
Giuseppe Conti, IT, Trilogis, iCore

Harold Liu, China, Wuxi Smart Sensing Stars, iCore
John Soldatos, GR, Athens Information Technology, OpenIoT
João Sarraipa, PT, UNINOVA, OSMOSE
Jose Antonio Galache, ES, Universidad de Cantabria, SmartSantander
Juan Ramón Santana, ES, Universidad de Cantabria, SmartSantander
June Sola, ES, Innovalia Association, FITMAN
Luis Muñoz, ES, Universidad de Cantabria, SmartSantander
Luis Sánchez, ES, Universidad de Cantabria, SmartSantander
Manfred Hauswirth, IE, INSIGHT Center, National University of Galway, OpenIoT
Marc Roelands, Alcatel Lucent, iCore
Martin Serrano, IE, INSIGHT Center, National University of Galway, OpenIoT
Maurizio Griva, IT, REPLY, OSMOSE
Maurizio Spirito, IT, ISMB, ALMANAC
Michele Sesana, IT, TXT e-solutions, FITMAN, OSMOSE
Michele Stecca, IT, M3S, iCore
Nikos Kefalakis, GR, Athens Information Technology, OpenIoT
Oscar Lazaro, ES, Innovalia Association, FITMAN
Pablo Sotres, ES, Universidad de Cantabria, SmartSantander
Pierluigi Petrali, IT, WHIRLPOOL, FITMAN
Raffaele Giaffreda, IT, CREATE-NET, iCore
Ricardo Goncalves, PT, UNINOVA, OSMOSE
Roberta Caso, IT, REPLY, OSMOSE
Sergio Gusmeroli, IT, TXT e-solutions, FITMAN, OSMOSE
Stephane Menoret, Thales, iCore
Stylianos Georgoulas, UK, University of Surrey, iCore
Swaytha Sasidharan, IT, CREATE-NET, iCore
Vera Stavroulaki, GR, UPRC, iCore
Veronica Gutierrez Polidura, ES, Universidad de Cantabria, SmartSantander

Contributing Projects and Initiatives

ALMANAC, OpenIoT, iCore, COMPOSE, SmartSantander, FITMAN, OSMOSE.

References

[1] McFedries, P., "The Cloud Is The Computer", IEEE *Spectrum*, August 2008.

[2] Vermesan, O. & Friess, P., "Internet of Things - Global Technological and Societal Trends", *The River Publishers Series in Communications*, May 2011, ISBN: 9788792329738.

[3] Soldatos, J., Serrano, M., and Hauswirth, M., "Convergence of Utility Computing with the Internet-of-Things", *International Workshop on Extending Seamlessly to the Internet of Things (esIoT)*, collocated at the *IMIS-2012 International Conference*, 4th6th July, 2012, Palermo, Italy.

[4] Hassan, M.M., Song B., Huh, E., "A framework of sensor-cloud integration opportunities and challenges", *ICUIMC* 2009, pp. 618–626.

[5] Lee K., "Extending Sensor Networks into the Cloud using Amazon Web Services", *IEEE International Conference on Networked Embedded Systems for Enterprise Applications* 2010, 25th November, 2010.

[6] Fox, G.C., Kamburugamuve, S., and Hartman, R., "Architecture and Measured Characteristics of a Cloud Based Internet of Things API", Workshop 13-IoT Internet of Things, Machine to Machine and Smart Services Applications (IoT 2012) at *The 2012 International Conference on Collaboration Technologies and Systems (CTS 2012)* May 21–25, 2012 The Westin Westminster Hotel Denver, Colorado, USA,

[7] Compton, M., Barnaghi, P., Bermudez, L., Castro, R.G., Corcho, O., Cox, S., et. al.: "The SSN Ontology of the Semantic Sensor Networks Incubator Group", *Journal of Web Semantics: Science, Services and Agents on the World Wide Web*, ISSN 1570–8268, Elsevier, 2012.

[8] Aberer, K., Hauswirth, M., Salehi, A., "Infrastructure for Data Processing in Large-Scale Interconnected Sensor Networks", *MDM 2007*, pp.198–205.

[9] Boehm, B., Abts, C., Brown, A.W., Chulani, S., Clark, B.K., Horowitz, E., Madachy, R., Reifer, D.J., and Steece, B., "Software Cost Estimation with COCOMO II", *Englewood Cliffs, NJ:Prentice-Hall*, 2000. ISBN 0-13-026692-2.

[10] Serrano, M., Hauswirth, M., Soldatos, J., "Design Principles for Utility-Driven Services and Cloud-Based Computing Modelling for the Internet of Things", *International Journal of Web and Grid Services* (Inderscience Publishers), ISSN: 1741-1106 (Print), 1741–1114 (Online) Subject: Computing Science, Applications and Software and Internet and Web Services, Volume 10, Number 2–3/2014.

[11] Petrolo, R., Mitton, N., Loscri, V., Soldatos, J., et. al., "Integrating Wireless Sensor Networks within a City Cloud", *Self-Organizing Wireless Access Networks for Smart City (SWANSITY), part of IEEE SECON 2014*, Singapore June 30th 2014.

[12] Guinard, D., Trifa, V., Mattern, F., Wilde, E., "From the Internet of Things to the Web of Things: Resource Oriented Architecture and Best Practices", In: Uckelmann, D., Harrison, M., Michahelles, F., (Eds.): *Architecting the Internet of Things*. Springer, ISBN 978-3-642-19156-5, pp. 97–129, New York Dordrecht Heidelberg London, 2011.

[13] Openstack, Open source software for Cloud services, http://www.openstack.org

[14] Cloudfoundry PaaS, http://www.cloudfoundry.com/

[15] CouchBase Server, http://www.couchbase.com

[16] Apache CouchDB, http://couchdb.apache.org

[17] Neumeyer, L., Robbins, B. ; Nair, A. ; Kesari, A., "S4: Distributed Stream Computing Platform", *2010 IEEE International Conference on Data Mining Workshops (ICDMW)*. Dec. 2010, pp. 170–177.

[18] STORM: Distributed and fault-tolerant realtime computation, http://storm-project.net/

[19] Apache Pig. http://pig.apache.org/

[20] Scala Framework, http://www.scala-lang.org/node/273.

[21] Sanchez, L., Muñoz, L., Galache, J.A., Sotres, P., Santana, J.R., Gutierrez, V., Ramdhany, R., Gluhak, A., Krco, S., Theodoridis, E., Pfisterer, D., "SmartSantander: IoTexperimentation over a smart city testbed", *Computer Networks*, 61 (2014), pp. 217–238

[22] SmartSantander project deliverable D1.3, http://www.smartsantander.eu/downloads/Deliverables/D13-Third-Cycle-Architecture-Specification-Final.pdf (accessed on 15th May 2014)

[23] FI-WARE initiative, http://www.fi-ware.org (accessed on 15th May 2014)

[24] SmartSantanderRA application, http://www.smartsantander.eu/index.php/blog/item/174-smartsantanderra-santander-augmented-reality-application (accessed on 15th May 2014)

[25] Pace of the City application, http://www.smartsantander.eu/index.php/blog/item/181-participatory-sensing-application (accessed on 15th May 2014)

[26] FITMAN (Future Internet Technologies for MANufacturing industries), http://www.fitman-fi.eu/

[27] Digital Business Community, http://www.dbi-community.eu/

[28] Keith Nosbusch, Cisco's Internet of Things World Forum 2013
[29] OSMOSE (OSMOsis applications for the Sensing Enterprise), http://www.osmose-project.eu
[30] OSMOSE user requirements, http://gris-dev.uninova.pt/osmose
[31] OpenIoT, http://www.github.com/OpenIotOrg/openiot

8

Bringing IP to Low-power Smart Objects: The Smart Parking Case in the *CALIPSO Project*

Paolo Medagliani,[1] Jérémie Leguay[1] Andrzej Duda,[2] Franck Rousseau[2]
Simon Duquennoy,[3] Shahid Raza[3] Gianluigi Ferrari,[4] Pietro Gonizzi,[4]
Simone Cirani,[4] Luca Veltri[4] Màrius Monton,[5] Marc Domingo,[5] Mischa
Dohler,[5] Ignasi Vilajosana[5] Olivier Dupont[6]

[1] *Thales Communications & Security, France*
[2] *Centre National de la Recherche Scientifique, France*
[3] *Swedish Institute of Computer Science, Sweden*
[4] *University of Parma, Italy*
[5] *Worldsensing, Spain*
[6] *Cisco System International, Netherlands*

Abstract

The chapter describes the Calipso communication architecture for IP connectivity in wireless sensor networks (WSNs) and the Smart Parking application scenario developed within the Project. The use case is a real life demonstrator for traffic flow and parking monitoring deployed in the city of Barcelona, Spain. It is based on a communication infrastructure with sensors for parking and traffic detection embedded in the ground. The sensor nodes communicate parking space availability/traffic flow to neighboring sensors until they reach a gateway. Multi-hop routing is used when there is no direct communication with the gateway. A centralized control system stores and processes all data gathered from sensors. The resulting information and implemented services are offered to citizens by means of mobile applications and city panels. In the chapter, we analyze the requirements of the use case, present the communication architecture of Calipso, and show how the Smart Parking application takes advantage of different modules within the architecture.

8.1 Introduction

8.1.1 Bringing IP to Energy-Constrained Devices

The Internet of Things (IoT) proposes the vision of interacting with the physical world by interconnecting objects with processing, communication, sensing, and actuating capabilities. The main IoT challenges include the integration of small Smart Objects having strong energy and processing constraints, large-scale interconnection of nodes through flexible and secure networking, as well as personalized interaction with the physical world and integration within the user-created content and applications.

Existing solutions stemming from past industrial and academic initiatives suffer from several limitations. The most important obstacle in the development of the Internet of Things was the advent of a large number of proprietary or semi-closed solutions such as Zigbee, Z-Wave, Xmesh, SmartMesh/TSMP that proposed different functionalities at several layers. Moreover, early research works in sensor networks suggested that the constrained and application-specific nature of sensor networks required networking to be based on non-IP concepts [8], [13]. The result was the existence of many non-interoperable solutions addressing specific problems and based on different architectures and different protocols. Such approaches have not led neither to large-scale deployments nor to interconnection of products from different vendors. Interconnecting heterogeneous networks is possible via protocol translation gateways, but this approach also presents several problems (reduced scalability, potential security issues, no end-to-end services, etc.).

For a long time, using IP in constrained networks was considered as too complex or ill-suited for such environments. However, with the increase of computing power and memory size, several successful implementations have showed the possibility of running the full-fledged IP stack [1], [2], [14], [15], [17], [22], showing that the performance of layered IP-based sensor network systems rivals that of ad-hoc solutions. One of the salient examples is the Contiki operating system, which was first used to explore IPv4 communications for sensor networks [2], [4] and later provided the first fully certified IPv6 stack for IP-based smart objects [7]. Hui and Culler have developed an IPv6 architecture for low-power sensor networks based on IEEE 802.15.4 [14].

Running an IP stack on a Smart Object presents the advantage of easy integration with the current Internet and an easy reuse of existing applications or protocols. More specifically, IP provides several important characteristics:

- it is based on open standards, which is essential for interoperability, cost efficiency and innovation
- intelligence is pushed outside the network, enabling not only network administrators but also users to develop new applications
- flexibility—supporting a wide range of media and devices
- universality—all protocols that solved very specific issues never survived
- open support for security
- support for auto configuration
- scalability.

Obviously, the minimal computing and memory requirements for running the protocol limit the all IP approach to objects that may currently cost about tens of euros. Smaller, less powerful nodes may still operate without IP to perform some specialized functions, if the cost justifies such a choice.

The Pervasive Internet needs a universal alternative to the many existing techniques for connecting ordinary devices to the Internet. The other techniques all have something to recommend them; each is optimized for a special purpose. But in return for their optimality, they sacrifice compatibility. Since most device connectivity rarely requires maximum optimality, compatibility is a much more important objective. IP is the only answer. End-to-end IP architectures are widely accepted as the only alternative available to support the design of scalable and efficient networks comprised of large numbers of communicating devices. IP enables interoperability at the network layer, but does not define a common application-layer standard, thus making it optimal for use in a wide variety of applications ranging across several industries [11].

8.1.2 The CALIPSO Project

CAL*I*PSO is a European FP7 project targeting the development of IP connected smart objects. In order to provide long lifetime and high interoperability, novel methods to attain very low power consumption are put in place. CAL*I*PSO leans on the significant body of work on sensor networks to integrate radio duty cycling and data-centric mechanisms into the IPv6 stack, something that existing work has not previously done. In the CAL*I*PSO project, we propose a number of enhancements to the standard low-power IP stack such as protocol optimizations, new network protocols, or security modules.

The context of CAL*I*PSO is the IETF/IPv6 framework, which includes the recent IETF RPL and CoAP protocols. It sets up a structure for evaluation that has not previously been available. Implementations have been carried out within the Contiki open source OS, the European leading smart object OS. In order to drive the development, three applications have been considered: Smart Infrastructures, Smart Parking, and Smart Toys, all of which need both standardized interfaces and extremely low power operation.

CALIPSO considers that smart object networks both need to communicate with other smart objects, other smart object networks, as well as Internet-based systems. The project goal is to push IP end-to-end connectivity all the way into smart objects through compact, energy efficient, and loss/failure tolerant routing and radio protocols. CALIPSO focuses on four specific layers of the IP stack: the MAC layer, the routing layer, the transport layer, and the application layer. With a deep understanding of the complex interactions between the layers, CALIPSO is able to significantly increase the performance and reduce the power consumption of IP-based smart object networks, thereby removing major barriers to IP adoption in smart object networks.

8.2 Smart Parking

One of the key applications and entry-points to Smart Cities is the Smart Parking application. It is designed to help drivers in the tough process of finding a parking spot in a crowded city. With the help of this kind of applications, citizens can reduce the searching time by 8% on the average, allowing them to save time, fuel, and associated costs, and hence, reducing frustration, accidents, and increasing the quality of life in cities. Because urban traffic is the cause of 40% of CO_2 and 70% of other contaminant in cities, Smart Parking applications also reduce overall city contamination.

Last but not least, deploying a Smart Parking application in a city with controlled parking areas, their occupation time is increased, the number of non-paying drivers drops and in conclusion, the total income of the municipality can be increased by almost 15%.

Some Smart Parking applications are based on cameras aiming at parking zones and streets with all the problems inherent to image processing applications (image quality, changing conditions, high bandwidth needs, etc.). The Smart Parking application specified in CALIPSO is based on individual car sensor devices installed at every single parking spot in the city. Every device processes the signal received by car sensing techniques and it decides if there is a car parked above or if it is a free spot and sends this information to a central server, where the data is processed, clustered, and sent to the citizens in various ways (mobile phone application, on-street panels, information website, etc.).

In parallel to that, municipality and city traffic control can retrieve and manage the current and historical data to study how to enhance traffic management in the city, adjust fees on controlled parking areas, etc. This data is of a great value to the city as long as the information forms a new axis on the city data space.

Figure 8.1 shows the communication infrastructure with the sensors for parking and traffic detection embedded in the ground. The sensor nodes communicate the parking space availability/traffic flow to neighboring sensors until the data reach the gateway. Multi-hop routing is used when the direct contact with the gateway cannot be made. A centralized control system stores and processes all the data gathered from sensors.

The storyline of the use case is the following: a driver heading to a desired parking sub-area (i.e., within a specified walking distance to the final destination) can use this service from the mobile phone. First, the system will tell if it forecasts that free parking spaces will be available at the expected time of arrival. The availability forecast will be done at the central platform using algorithms based on context information (time of the day, day of the week, weather conditions, etc.) and historic data. Two outcomes are possible:

Case 1: available parking spaces are forecasted in this sub-area. In this case, the system will advise parking spaces with the largest numbers of free spots. Also, the system will notify the driver if traffic congestion has been detected along the route up to the selected parking space.

Case 2: no available parking spaces are forecasted in this sub-area. In this case, the system will search and recommend another sub-area with available

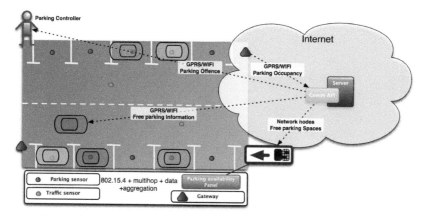

Figure 8.1   Architecture of the parking space availability control service

parking spots according to user preferences. The preferences will include the proximity to the desired area and traffic status along the route.

The use case 1 will demonstrate the parking control application that consists of the following main components:

- Sensor Nodes, which are small-embedded devices containing an AMR (Anisotropic Magneto-Resistive MEMS) sensor, signal conditioning stages through FPAA, a low-power IEEE 802.15.4 wireless interface for communication. These nodes are connected to a self-organizing network for communication between nodes.
- Hybrid Gateways collecting information about parking availability from sensor nodes on the streets and transmitting the information to the centralized urban control through the Internet. They allow interconnection using different interfaces in order to be easily adapted to different urban scenarios like urban WiFi, wired municipality infrastructure, fiber optical, etc.
- Cloud Central Platform collecting the information sent by the gateways and city sources, and implementing the service to be accessed by the final users (through Information Panels and Mobile Phones). This service will identify the available parking spots and offer a forecast.
- Information Panels collecting parking availability information from the control center and display this information to guide citizens to find free parking spots.
- Mobile application to run on the users portable devices to access the information in real time and obtain recommendations.

In Figure 8.2 we show an example of real-time panel information displaying and control interface used in Smart Parking applications (the Worldsensing Smart Parking application were deployed in Barcelona and Moscow).

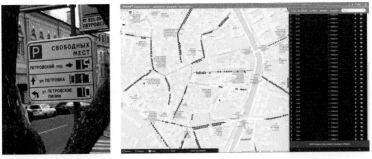

Figure 8.2   Installation and control interface of a Smart Parking application

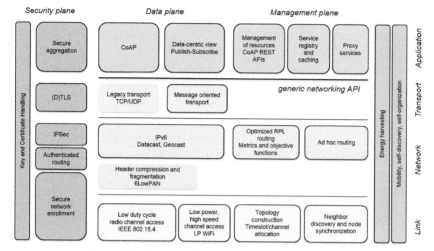

Figure 8.3 High level functional architecture of sensors

Table 8.2 summarize the requirements of the Smart Parking application developed by Worldsensing SME. The requirements are based on market demands and user experience like a long battery life (about 5 years), maximum delay (i.e. response time of the application must be less than 10 seconds), etc.

8.3 CALIPSO Architecture

This section presents the CAL*I*PSO architecture. Figure 8.3 shows a layered view of the architecture for a Smart Object node. Bold shapes indicate the blocks on which the project focused by providing enhancements, optimizations, or adaptations to Smart Objects constraints.

In this figure, we do not present the gateway (or LBR, the border router in the IETF terminology) that interconnects the constrained network with the Internet. This element supports the standard TCP/IP protocol stack and adapts its operation to Smart Object nodes. The main function of the gateway relates to the 6LoWPAN layer that takes care of fragmenting IPv6 packets longer that the L2 MTU and compresses headers. It can also provide interfacing CoAP with the standard HTTP.

At PHY/MAC layer, the constrained nodes should benefit from energy efficient solutions such as IEEE 802.15.4 or Low Power WiFi, depending on application needs for bandwidth and delay. On the other side, the gateway must be able to communicate both with nodes that is running the same PHY/MAC protocols, and with the Internet, running standard Ethernet and WiFi protocols.

Similar considerations can be carried out for the networking layer. The constrained nodes run the IPv6 protocol, coupled to 6LoWPAN to adapt to the underlying data link layer. The routing protocols and the data communication paradigms as well are specific to the constrained domain. The gateway, instead, in addition to the above mentioned solutions, must also run traditional legacy IPv4 and IPv6 to interconnect to existing network infrastructures.

At the transport layer, both gateway and nodes can run legacy protocols such as TCP or UDP, with a preference for UDP in the constrained world, due to its reduced memory and resource use. In addition, nodes can implement a publish/subscribe mechanism to improve data collection and reduce unneeded communications.

Finally, at the application layer, nodes run specific lightweight protocols to enable efficient communications, such as caching of data or REST-like

Table 8.1   Requirements of the Smart Parking

Aspect	Requirement
Physical/link layer	802.15.4 with a duty-cycling scheme
Topology	Multi-hop network
Throughput and latency	Low throughput (collection of periodic values every 4s). Support for bursty load (car footprint transmission).
Energy consumption	Battery-powered device. Lifetime required for 5 years
Duty cycling MAC	Yes. Tailored for convergecast traffic (Multipoint-to-Point) for Smart Parking data
Data aggregation/storage and in-network processing	Both required, data aggregation and in-network processing
Support of mobility	No
Traffic patterns	Multi-hop convergecast
Routing	Fairness and load balancing to mitigate hot spot problems. Evaluate metrics for RPL as Smart Parking application suffers from extreme multipath that requires smart updates on the routing topology.
Transport	QoS at transport
Neighbour and service discovery	Self-discovery of capabilities provided by nodes, announced/pulled out by the gateways, and/or in the vicinity.
Security	Payload and header encryption

interfaces adapted to the constrained world. The gateway, in addition to run the same protocols, is in charge of the interconnection with the Internet, therefore it exposes REST APIs that allow remote users to easily interact with the gateway.

Gateways do not have constraints in terms of energy and computational capabilities and this kind of issues primarily concerns nodes that need to deal with energy saving, self-organization, mobility, and security. Nodes appearing in the network must be able to automatically learn about network parameters. Eventually, the running protocols must take into account that nodes can move and react to the change of a position. Securing communications is another important aspect of the constrained world since the computational capabilities do not allow for traditional security mechanisms. Since nodes are battery powered or energy harvested, saving energy becomes crucial to extend the whole network lifetime. These aspects have an impact on the whole architecture of a node, requiring a careful specific design of the running protocols.

In the following, we detail the most important project contributions included in the CAL*I*PSO stack, as well as some existing protocols and solutions that have been exploited within CAL*I*PSO.

Table 8.2 Low level details of the Smart Parking

Application	
Maximum time since a car arrive until it is shown on the display	10 sec
Gateway	
Maximum nodes per gateway	> 50
Motes	
Maximum transmission distance between motes (car on)	Each device must reach 2 other devices, within 10 meters.
Maximum transmission distance between motes (no car on)	Each device must reach 3 other devices, within 15∼20 meters.
Sampling rate	1 Hz
Radio	
Size of packets	20 bytes
Number of packets per minute	Each time a car comes in/out + keep-alive every 15 minutes
Duty cycle	$<0.2\%$
Percentage of lost messages	$<10\%$

8.3.1 CALIPSO Communication Modules

8.3.1.1 MAC layer
RAWMAC

RAWMAC is a cross-layer mechanism in which the routing layer, for instance the RPL protocol presented in Section 8.4.1.2, is used as a management layer for organizing the asynchronous duty-cycled MAC protocols (such as ContikiMAC).

ContikiMAC [3], [6] is the Contiki default low-power listening MAC protocol. To listen, nodes periodically wake up (e.g. at 8 Hz) and proceed with two Clear Channel Assessments (CCAs) with an interval that guarantees that any ongoing packet transmission will be sensed. Upon detecting activity on the channel, the node keeps its radio on, waiting for an incoming packet. To send, nodes transmit the data packet repeatedly as a wakeup signal, and keep doing so until they received an acknowledgement (case of unicast) or for exactly one wakeup period (case of broadcast).

ContikiMAC implements a so-called "phase-lock" optimization where senders remember the wakeup phase of each of their neighbors to optimize subsequent transmissions (wakeup signal starts briefly before expected target wakeup). Once a receiver is active, the transmitter can transmit arbitrarily long sequences of packets to amortize the wakeup procedure and allows for efficient forwarding of large data bursts [6]. The contention-based, unscheduled nature of ContikiMAC allows it to handle random traffic patterns while sleeping more than 99% of the time. Because it emulates an always-on link and makes no assumptions on the above layers, ContikiMAC is an ideal choice for low-power IP scenarios.

The key idea of RAWMAC is to exploit the DODAG built by RPL to make each node align its wake-up phase with that of its preferred parent, creating a data propagation "wave" from the leaves of the DODAG to the root. Once a data packet rides this wave, the latency is significantly reduced, as it depends only on small propagation delays and on the internal processing carried out at each device to forward the packet. By properly configuring the phase lock mechanism of ContikiMAC, the transmitting node wakes up only when the receiving node is ready to receive the packet, so that the energy consumption is kept as low as possible.

Figure 8.4 shows this wake-up phase alignment in RAWMAC. As long as the routing structure is established, a node shifts its wake-up phase in order to be aligned with that of its parent. More precisely, it sets the wake-up phase to the time at which it received the last link layer ACK from its RPL

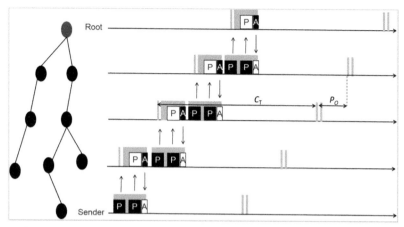

Figure 8.4 Principle of RAWMAC: the wake-up phase alignment. Once the RPL DODAG is set up, each node shifts its wake-up phase according to that of its parent in the DODAG

preferred parent. Since the preferred parent must have been awake to receive the packet, the node can assume that the reception of the ACK means that it has successfully transmitted a packet within the preferred parent wake-up window and, thus, that it has found the preferred parent wake-up phase.

We define the phase offset P_o (dimension: [s]) as the offset between the node wake-up phase and the wake-up phase of its parent. Given the node sleeping interval C_T, it holds that $0 \leq P_o \leq C_T$. The parameter Po has indeed to be chosen carefully, since it has an impact on the system delay performance. If P_o is too short, a node relaying a packet may not be able to catch its parent wake-up, because the reception of the same packet from its child has not completed yet. If this is the case, then the child should wait the next cycle time C_T to be able to forward the packet. If P_o is too long, instead, the delay significantly increases, as the sender has to wait for the receiver to wake up to be able to transmit the packet.

8.3.1.2 Routing layer
RPL

The RPL routing protocol [9], [21] has been developed for a limited data rate and lossy environments. Actually, RPL is the most adopted routing protocol for constrained networks. RPL is a distance-vector protocol based on the creation of a routing topology referred to as the *Destination Oriented Acyclic Directed Graph* (DODAG), in which the cost of each path is evaluated according metrics defined in an objective function. The goal of this protocol is the creation of a collection tree protocol, as well as a point-to-multipoint network from the root of the network to the devices inside the network.

To keep the status of the network updated, the root of the RPL DODAG periodically broadcasts *DODAG Information Object* (DIO) control messages. The receiving nodes may relay this message or just process it, if configured as leaves of the tree. The RPL protocol also introduces a *trickle* mechanism that allows reducing the transmission frequency of DIO messages according to the stability of the network.

In some cases, when specific flags in DIO packets are set, the nodes receiving a DIO are stimulated for the generation of a *Destination Advertisement Object* (DAO) messages. This is a unicast data packet that can be sent either directly to the root, when a *non-storing* mode is used or to the selected parent in the *storing* mode. The messages create downstream routes from the root to the leaves.

In the former case (non-storing mode), intermediate nodes simply add their addresses to the DAO header. Only the root stores the downward routing table of the tree. In the latter case (storing mode) instead, since DAO messages are directly processed by the parent node that receives the packets, each node stores a routing table for the children associated to it. Before sending in its turn a DAO, a node aggregates the information received by the children, so that the aggregated reachability information is sent to its parent. As unicast messages, the DAO can be acknowledged by the receiver.

The third type of message foreseen by RPL is the *DODAG Information Solicitation* (DIS) used by nodes to advertise their presence in the network so that they can join an existing DODAG.

ORPL

ORPL [5] is an opportunistic extension to RPL. The basic idea behind ORPL is to replace unicast transmissions to a specific next hop by anycast transmissions aimed at any node that offers progress towards the destination. Figure 8.5 illustrates the anycast operation in which traditional routing estimates link quality and sticks to links that appear to be generally good, while ORPL uses any link available at the time of transmission.

Combined with ContikiMAC radio duty cycling, ORPL conciliates low energy (nodes sleep most of the time) and low latency (first awoken neighbor forwards).

Routing in ORPL is possible towards the root by simply following a gradient, but also towards any other node by going away from the gradient, directed by lightweight routing tables that merely contain the set of nodes below in the topology ("routing sets").

ORPL was tested at a large scale and has shown to attain delivery ratios over 99% together with sub-percent duty cycles and sub-second latency.

RRPL

Reactive RPL is a lightweight version of RPL that retains the collection tree structure (DODAG) of RPL, but speeds up local repair in case of link failures and allows for reactive or proactive routing for downward (P2MP) traffic. It provides a mechanism for fast local route repair through the use of a link reversal algorithm towards the sink. It takes advantage of the existing DODAG structure to enable efficient reactive route search. Moreover, RRPL does not impose the use of the additional RPL header in each packet, since it quickly detects and fixes routing loops.

Featurecast

Featurecast is a new address-centric communication paradigm especially well-suited for sensor networks. It uses a data-centric approach to select destination nodes: destination addresses correspond to a set of features characterizing sensor nodes. For instance, we can reach a group of nodes satisfying the

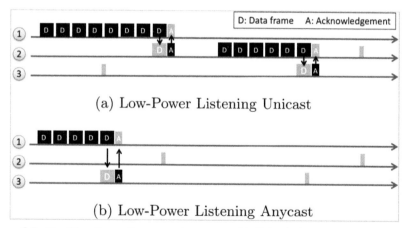

(a) Low-Power Listening Unicast

(b) Low-Power Listening Anycast

Figure 8.5 Traditionally, routing uses unicast over stable links aiming at stability. Low-power listening introduces a significant delay at every hop. ORPL uses anycast and transmits to the first awoken forwarder that receives the packet regardless of link quality estimates

featurecast address [4th floor and temperature]. In this way, the communication mode closely reflects how application developers reason about sensors, actuators, and their interaction with the real world. Features may be freely defined and specific to a particular application or a given deployed network.

Featurecast extends the standard notion of multicast with a more general definition of groups: instead of one address representing a multicast group, a featurecast address defines the group membership based on a set of features. With featurecast, a node has implicitly an address for every subset of its advertised features. We propose a scheme for efficiently routing packets to all such addresses as well as an address coding compatible with the multicast IPv6 address format.

8.3.1.3 Application layer
CoAP
RESTful web services and the HTTP protocol are widely used to publish the status of resources. However, web services are not suitable for constrained networks due to the high overhead introduced by HTTP and the presence of the TCP congestion window. For these reasons, the CoRE IETF working group aimed at adapting the REST architecture to constrained networks through the definition of a protocol referred to as Constrained Application Protocol (CoAP) [20]. The RESTful architecture, as the idealized model of the Web, is usually implemented with HTTP, and its 4 basic methods: GET, POST, PUT, and DELETE.

In CoAP, the same four REST verbs have been implemented for constrained devices. CoAP is by nature more lightweight than HTTP, supports multicast natively, as well as the publish-subscribe model. It is for example perfectly possible with CoAP to subscribe with a single request to all smoke detectors in a building via a multicast request. In case of fire, an alarm can then be multicasted to all subscribers. CoAP supports reliable communication as an option on top of UDP. Actually, CoAP has become a de facto standard to expose web services in constrained networks.

HTTP/CoAP Proxy
The HTTP/CoAP (HC) proxy with caching functionality has been implemented relying on the IETF protocols. We made this choice both because some libraries, which can be reused to speed up the implementation, are already available and because IETF protocols have been specifically designed to meet the constraints typical of IoT architectures.

In Figure 8.6, we show a logical scheme of the implemented caching system, where the existing CoAP servers expose some resources to be queried. The running routing protocol is RPL. All the requested information and the notifications from the nodes are gathered to the root of the RPL tree, which overlaps with the WSN gateway. The HC proxy will then send the requested information to the final user.

Our HC proxy implementation is based on the Californium open source framework [16]. The main reason of this choice is that Californium already implements in JAVA a basic set of CoAP functionalities. We have then introduced two information storing mechanisms. The first one is the caching database, where all the information from the WSN are gathered and stored and for which we have chosen CouchDB, since it offers a REST HTTP API to query stored results. The second one is a subscription register to efficiently manage the subscribers to CoAP resources.

Requests are intercepted by the HC proxy that also handles the eventual response from a given CoAP server. If the proxy has a stored value that is fresh enough, that is whose lifetime is smaller than a given value, it directly replies to the request of a remote client, without forwarding it into the WSN. Otherwise, if the required value is not present or it is too old, it transfers the request to the intended CoAP server. Additionally, the proxy stores the sensor responses in the cache to make them available for other eventual incoming requests. A similar approach is used for the publish-subscribe register. In the case of observation requests issued by a remote client, the proxy handles them by maintaining a list of observed resources and a list of interested clients. Each time a notification for a resource update is sent from the node to the proxy, the proxy will forward this message to all interested subscribers.

Caching and observe mechanisms can operate together as well. For instance, the information already requested by an observe request can be

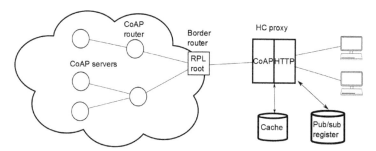

Figure 8.6 Scheme of the implemented HTTP/CoAP caching system

cached and made available to another request not related to the previous observation. In fact, the caching and the observation mechanism are independent from each other. Nevertheless, the two systems should be used together because the information obtained by observation can help the caching operations.

8.3.2 CALIPSO Security Modules

Compression of IPsec AH and ESP Headers for Constrained Environments

The security in 6LoWPAN networks is particularly important as we connect smart objects to the insecure Internet. The standardized and mandatory security solution for IPv6 is IPsec. We have proposed an extension of 6LoWPAN defining header compression for IPsec datagrams [18–19] .

We focus on the IPsec transport mode that provides end-to-end security, both from device to device and from device to traditional Internet hosts.

Our solution supports both the AH and ESP protocols, where:

- AH authenticates the IPv6 header and payload between two end hosts;
- ESP authenticates and encrypts the IPv6 payload (but not the header) between the end hosts.

When needed, IPsec AH and ESP can be combined to link-layer security to encrypt and authenticate the entire payload of the frame in a hop-by-hop fashion.

Our solution brings standard Internet-class security to the most constrained devices, i.e. devices running Contiki on 16-bit MCUs and with only tens of kilo-bytes of memory [19].

Distributed key verification and management

Distributed key certification renders sensor nodes fully autonomous and does not require the use of central authorities and certificates. We have designed a protocol based on asymmetric cryptography and one-way accumulators. It provides secure node enrollment and key certification suitable for other security protocols like IPsec and DTLS.

The protocol relies on Elliptic Curve Cryptography (ECC) for public/private keys as well as the accumulator. Each node is assigned a pair of keys and the material needed for the one-way accumulator: its witness, and the accumulator containing all the public keys of the nodes in the network, including the gateway. Two nodes can then assess that they are part of the same network by sending their public key and the witness to each other for mutual verification. Once the nodes have mutually verified their public keys, they can establish a symmetric key through the Elliptic Curve Diffie-Hellman (ECDH) key agreement protocol.

Our protocol inherits the major security properties of one-way accumulators: "one-way-ness" and resistance to forgery. Regarding node injection, the security of our protocol is reduced to the security of the accumulator. Due to the one-way-ness of accumulators, the capture of a node does not compromise the communications of other nodes: only the keys of the captured node are compromised. Finally, we deal with denial-of-service by triggering the most expensive operation (ECDH) using a table lookup, hence preventing the replay of correct messages that would cause the exhaustion of the node resources otherwise.

OAuth

Open Authorization (OAuth) [10] is an open protocol that allows secure authorization in a simple and standardized way from third-party applications accessing online services, based on the REST web architecture. OAuth has been designed to provide an authorization layer, typically on top of a secure transport layer such as HTTPS.
OAuth defines three main roles:

- the User (U) is the entity who generates some sort of information;
- the Service Provider (SP) hosts the information generated by the users and makes it available through APIs;
- the Service Consumer (SC), also referred to as "client application", accesses the information stored by the SP for its aims.

In order to allow a client application access information on his/her/its behalf, a user must issue an explicit agreement. The agreement results in the grant of an access token, containing the user's and client application's identities. The client must exhibit the access token in every request as an authorization proof. The OAuth 2.0 protocol enhances the original OAuth protocol focusing on the easiness of client development [12].

Smart objects providing CoAP-based services might also require some authorization for permitting access by third-party applications. In order to meet this kind of security requirement, smart objects might benefit by the use of the OAuth protocol. However, due to the need to execute heavy cryptographic computation and memory footprint issues (both in terms of available ROM and RAM on smart objects), it is not feasible to implement the OAuth logic and access-token management directly on the device.

For this reason, a novel OAuth-based authorization framework targeted to IoT scenarios has been designed and implemented. The authorization framework, denoted as "IoT-OAS", allows smart objects to delegate authorization-related operations in order to minimize the memory occupation due to the implementation of specific software and storage modules.

The procedure takes an incoming OAuth-secured request and asks the IoT-OAS authorization service to verify the access token included in the request against a set of client and user credentials it stores, by using the appropriate digital signature verification scheme, as specified in the OAuth protocol definition (either PLAINTEXT with secure transport, HMAC, or RSA). Upon reception of the request, the IoT-OAS service computes the digital signature for the incoming message and performs a lookup in its internal credential store to see if the request matches client identity, user grants, and requested resource access. If the signature is verified and the resource is set to be accessible, the IoT-OAS service replies with a success response and the request can then be served. Otherwise, the request is blocked and a client error response is sent back to notify that the client is not authorized to access the requested resource.

The delegation of the authorization functionalities to an external service, which may be invoked by any subscribed host or thing, affects:

- the time required to build new OAuth-protected online services, thus letting developers focus on service logic rather than on security and authorization issues;
- the simplicity of the Smart Object, which does not need to implement any authorization logic but must only invoke securely the authorization service in order to decide whether to serve an incoming request or not;
- the possibility to configure the access control policies (dynamically and remotely) that the smart object (acting as SP) is willing to enforce, especially in those scenarios where it is hardly possible to intervene directly on the device.

The IoT-OAS architecture supports HTTP (secured through TLS transport) and CoAP (secured through DTLS transport) requests sent by external clients targeting services provided by smart objects, thus enabling the possibility to perform access control either on the device (before serving requests) or on HTTP/CoAP proxies at the border of a constrained network (filtering incoming requests, which are consequently delivered or not to the target smart object).

8.4 Calipso Implementation and Experimentation with Smart Parking

This section presents the implementation of the Calipso modules, described in the previous section, and the experimentation plan for the Smart Parking application.

8.4.1 Implementation of Calipso Modules

Figure 8.7 shows the integration of the different modules. As Calipso targets several IoT-based applications, the integration of the modules is made as flexible as possible, to allow the choice between different protocols depending on the deployed application.

At the MAC layer, we can choose between two different MAC protocols: (i) RAWMAC and (ii) ContikiMAC. RAWMAC is suitable for efficient data collection (i.e., mostly mono-directional traffic). In order to create a "wave", this protocol requires a routing layer that does not vary routing tables very often. For instance, it can rely on RPL since routes are created and maintained

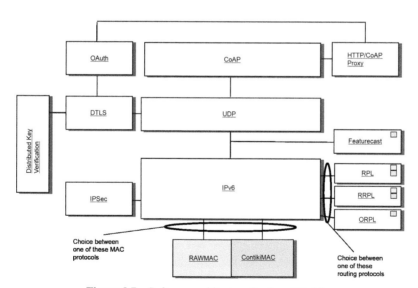

Figure 8.7   Software architecture for Smart Parking

proactively during time. On the other side, ContikiMAC is suitable in scenarios where low power communications are mostly bi-directional. Differently from RAWMAC, it does not require any information from the upper layer. Clearly, the choice of a specific MAC layer has an impact on the routing plane.

At the networking layer, we have the choice between three routing protocols: (i) ORPL, (ii) Reactive RPL, and (iii) RPL. ORPL is an extension of RPL where packets are forwarded opportunistically towards their destination, leading to increased reliability, shorter delay and reduced energy consumption. ORPL relies on anycast at the MAC layer (ContikiMAC), making it incompatible with the unicast-oriented RawMAC. ORPL performs at best in static, dense network, and supports random channel access. In use cases where only a few nodes are connected to the network at the same time and the impact of opportunistic strategies would be limited, it is preferable to adopt more conservative mechanisms providing anyway interesting capabilities such as fault tolerance and slow mobility management. RRPL instead, provides on-demand route recalculation. This allows to handle node (limited) mobility and to react to network failures. RRPL can coexist with RAWMAC since, as soon as a route is recalculated, RAWMAC follows route modification.

All the MAC and routing protocols presented above are perfectly compatible with IPv6, as well as with the IPSec security mechanism that allows providing secured communications. Featurecast instead is only compatible with ContikiMAC. In fact, Featurecast has the objective of providing an alternative to group communications for networks where no or limited mobility is provided. Since the considered routing protocols address unicast communications, whereas Featurecast is developed for multicast communications, these protocols can coexist natively. In contrast, Featurecast is not perfectly compatible with RAWMAC as the latter has been designed for aligning activity phases for efficient many-to-one communications. In Featurecast, instead, communications follow a many-to-many pattern. The two protocols could then coexist, but the delay introduced by RAWMAC would be significant.

At the transport layer, UDP is compatible with all the above-mentioned protocols, due to its stateless operation. In order to provide end-to-end security over UDP, we use DTLS. Since we want to reduce the overhead of cryptographic computation, we introduce a distributed key validation mechanism. In addition, as it could be important to authenticate users accessing the network, we use the OAuth solution. Because an authentication server cannot be executed directly on nodes for a matter of resource consumption, it has been integrated with the proxy node.

At the application layer we choose CoAP, which offers RESTful primitives to constrained networks. The most used verbs of CoAP are GET and Observe. In the former case, we deal with bi-directional communications, whereas in the latter, we mainly have mono-directional data transfer, matching very well a protocol for data collection such as RAWMAC.

When many requests from external users are carried out on the same node, it would be more efficient to have a caching node outside the constrained network replying on behalf of the in-network node. In addition, since requests from the Internet are coming in HTTP, it necessary to translate request (and responses) from HTTP to CoAP (and from CoAP to HTTP). The HTTP/CoAP proxy is then in charge of carrying out this protocol translation. The proxy is also able to publish data on remote webservers in case this is required by the use case.

8.4.2 Experimentation Plan for Smart Parking

The integrated modules will be evaluated in the Smart Parking application, whose requirements have been already defined in previous sections.

8.4.2.1 Prototype description

In the experiments, it is planned to use the Tmote Sky platform. This mote is an ultra-low power wireless module that leverages industry standards like USB and IEEE 802.15.4 to interoperate with other devices. The Tmote Sky has been the base for most of the developments done in Calipso.

The Tmote Sky has the following hardware characteristics:

- 250kbps 2.4GHz IEEE 802.15.4 Chipcon Wireless Transceiver
- 8MHz Texas Instruments MSP430 microcontroller (10k RAM, 48k Flash)
- Integrated onboard antenna with 50m range indoors / 125m range outdoors
- Integrated Humidity, Temperature, and Light sensors
- Ultra-low current consumption
- Programming and data collection via USB
- optional SMA antenna connector
- Contiki support

The Tmote is connected to a standard battery (see Figure 8.8 (a)) and placed into a Worldsensing box to be buried into the tarmac (see Figure 8.8 (b)).

(a)   (b)

Figure 8.8   Tmote with battery ready for boxing, and **(b)** Tmote in the Worldsensing box in the real deployment

8.4.2.2 Description of the scenario

Each device will be integrated with the Worldsensing box in order to be buried in tarmac in the same way a real installation is done. This installation procedure will mimic the exact conditions for a real deployment.

The test facility is situated in a corner street in Barcelona in the 22@ neighborhood. Its location can be seen in Figure 8.9. The corner is Parking Load Zone that it is restricted for loading and unloading activities, with a maximum stop time of 30 minutes. Each device is buried about 2 meters a part of each other, for a total of 6 devices to cover the entire corner.

This zone is selected due to its high car churn (car replacement) and its lack of marked parking space. This lack of markings causes different placement of any truck or car parking on the zone, changing the radio link between motes each time a car enters or leave the zone. These changes will be reflected on changes on the routing and the topology of the network created.

The network topology would be available through tools consulting the base station (Figure 8.10).

In order to test RAWMAC performance, nodes are configured in a chain topology as shown in Figure 8.10. As soon as a car is moved, a message is sent to the web application. We will measure the delay of transmission and the packet delivery ratio. In addition, we will fix a threshold on delay and we will compare the energy consumption of RAWMAC and ContikiMAC. We will also measure the convergence time while RRPL is establishing new routes, the lengths of the established routes/paths, as well as the traffic overhead of RRPL signaling.

Figure 8.9   Picture of the selected Smart Parking test facility

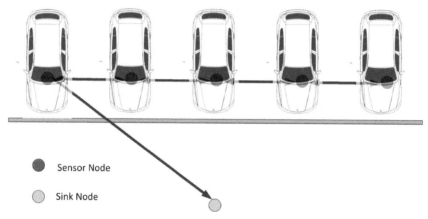

Figure 8.10   Configuration of the network topology

8.4.2.3 Performance indicators

For each of the tested modules, we provide the performance indicators that will characterize the operation of the proposed mechanisms.

RAWMAC

The performance of RAWMAC will be measured in terms of the following metrics:

- Packet delivery ratio: the ratio of packets successfully delivered to the base station.
- Energy consumption of the sensors.
- Upward delay (Optional): delay of a packet sent from a sensor to the base station.

Reactive RPL

The performance of reactive RPL will be measured in terms of the following metrics:

- Code footprint: the amount of flash memory the routing protocol uses when compiled into the node firmware.
- Convergence time and route length measure how quickly the routes are established and maintained as well as their length.
- Traffic overhead to maintain connectivity. Routing packets consume radio resources, but also computation time at the sender and receiver.

ORPL

The performance of ORPL will be measured in terms of the following metrics:

- Packet delivery ratio: the ratio of packets successfully delivered to the base station.
- End-to-end delay: delay from a node transmitting to the base station receiving it.
- Duty cycle: proportion of time with radio turned on, used as a proxy for power.
- Hop count: the number of hops for transmitted packets to reach the root (this metric does not reflect the end goal, but is interesting for network monitoring).

8.5 Concluding Remarks

In this chapter, we have first presented the CALIPSO communication architecture and the Smart Parking use case. Then, we have detailed the most relevant modules developed in CALIPSO showing how they contribute to meet the use case objectives and which interfaces are necessary to make the modules interoperable.

The example application clearly shows how IP can foster communications among constrained smart objects. Currently, the CALIPSO protocol stack is being tested and validated in the Smart Parking application. Our preliminary results show significant improvements in performance and energy consumption while extending IP reachability to small constrained devices. Other applications may benefit from the functionalities to bring new innovative services to citizens.

Future research extensions of CAL*I*PSO project will encompass the inclusion of energy harvesting techniques into the developed stack and the use of heterogeneous networks. In order to let the devices recharge, even more severe duty cycles at the nodes must be used. Thus, the solutions developed must be adapted to take into account this additional functionality.

Actual solutions for constrained devices barely allow the transmission of large amount of data due to the limited bandwidth offered, so coupling a high speed and high energy-consumption interface with a low power and low bandwidth interface will boost the overall network capacity.

Acknowledgements

The work of the authors is funded by the European Community's Seventh Framework Program, area "Internetconnected Objects", under Grant no. 288879, CAL*I*PSO project - Connect All IP-based Smart Objects. The work reflects only the authors' views; the European Community is not liable for any use that may be made of the information contained herein.

References

[1] Dawson-Haggerty, S., Jiang, X., Tolle, G., Ortiz, J., & Culler, D., "SMAP: a simple measurement and actuation profile for physical information". *Proceedings of the 8th ACM Conference on Embedded Networked Sensor Systems,* 2010, pp.197–210.

[2] Dunkels, A., "Full TCP/IP for 8-bit architectures". *Proceedings of The International Conference on Mobile Systems, Applications, and Services (MobiSys),* 2003, 14 pages.

[3] Dunkels, A., "The ContikiMAC Radio Duty Cycling Protocol", *Technical Report T2011:13.* Swedish Institute of Computer Science, December 2011.

[4] Dunkels, A., Voigt, T., & Alonso, J., "Making TCP/IP Viable for Wireless Sensor Networks", *Proceedings of the European Conference on Wireless Sensor Networks (EWSN),* 2004, 4 pages.

[5] Duquennoy, S., Landsiedel, O., & Voigt, T., "Let the Tree Bloom: Scalable Opportunistic Routing with ORPL", *Proceedings of the International Conference on Embedded Networked Sensor Systems (ACM SenSys 2013)*, Rome, Italy, November 2013.

[6] Duquennoy, S., Österlind, F., & Dunkels, A., "Lossy Links, Low Power, High Throughput", *Proceedings of the International Conference on Embedded Networked Sensor Systems (ACM SenSys 2011)*. Seattle, WA, USA, November 2011.

[7] Durvy, M. et al., "Making Sensor Networks IPv6 Ready", *Proceedings of the International Conference on Embedded Networked Sensor Systems (ACM SenSys)*, 2008, pp. 421–422.

[8] Estrin, J. H., Govindan, R., & Kumar, S., "Next century challenges: scalable coordination in sensor networks", *Proceedings of the fifth annual ACM/IEEE international conference on Mobile computing and networking*, 1999, pp. 263–270.

[9] Gaddour, O., & Koubâa, A., "RPL in a nutshell: A survey", *Computer Networks: The International Journal of Computer and Telecommunications Networking*, 2012, pp. 3163–3178.

[10] Hammer-Lahav, E., "The OAuth 1.0 Protocol - RFC 5849", April 2010.

[11] Harbor Research. M2M & Smart Systems, Machine-To-Machine (M2M) & Smart Systems. London: Harbor Research, 2011.

[12] Hardt, D., "The OAuth 2.0 Authorization Framework - RFC 6749", October 2012.

[13] Hill, J., "System Architecture Directions for Networked Sensors", *ACM SIGOPS Operating Systems Review*, 2000, pp. 93–104.

[14] Hui, J., & Culler, D., "IP is Dead, Long Live IP for Wireless Sensor Networks", *Proceedings of the International Conference on Embedded Networked Sensor Systems (ACM SenSys)*, 2008, 14 pages.

[15] Jiang, X., Dawson-Haggerty, S., Dutta, P., & Culler, D., "Design and implementation of a high-fidelity ac metering network", *Proceedings of the International Conference on Information Processing in Sensor Networks (ACM/IEEE IPSN)*, 2009, 12 pages.

[16] Kovatsch, M., Mayer, S., & Ostermaier, B., "Moving Application Logic from the Firmware to the Cloud: Towards the Thin Server Architecture for the Internet of Things", *Proc of the 6th Int Conf on Innovative Mobile and Internet Services in Ubiquitous Computing (IMIS 2012)*, 2012, 6 pages.

[17] Priyantha, B., Kansal, A., Goraczko, M., & Zhao, F., "Tiny web services: design and implementation of interoperable and evolvable sensor networks", *Proceedings of the International Conference on Embedded Networked Sensor Systems (ACM SenSys)*, 2008, pp. 253–266

[18] Raza, S., Duquennoy, S., & Selander, G., "Compression of IPsec AH and ESP Headers for Constrained Environments", 2014, Retrieved from http://tools.ietf.org/html/draft-raza-6lo-ipsec-00

[19] Raza, S., Duquennoy, S., Chung, T., Yazar, D., Voigt, T., & Roedig, U., "Securing Communication in 6LoWPAN with Compressed IPsec", *7th IEEE International Conference on Distributed Computing in Sensor Systems (DCOSS '11)*. Barcelona, Spain, June 2011.

[20] Shelby, Z. et al., "Constrained Application Protocol (CoAP)", *draft-ietf-core-coap-18*, 2013.

[21] Winter, T. et al., "RPL: IPv6 Routing Protocol for Low-Power and Lossy Networks. RFC 6550", March 2012.

[22] Yazar, D., & Dunkels, A., "Efficient Application Integration in IP-Based Sensor Networks", *Proceedings of the ACM BuildSys 2009 workshop, in conjuction with ACM SenSys 2009*, 6 pages.

9

Insights on Federated Cloud Service Management and the Internet of Things

Martin Serrano,[1] Manfred Hauswirth,[1] John Soldatos,[2] Nikos Kefalakis[2]

[1]*Digital Enterprise Research Institute, DERI- NUIG, Galway, Ireland*
[2]*Research and Education Laboratory in Information Technologies, AIT, Peania, Greece*

Abstract

The current evolution towards using cloud-based infrastructures in the provisioning of Internet of Things (IoT) services is creating an inherent association between the Internet-connected devices, the stored and collected data (from the IoT devices) and the way in how the cloud infrastructure is managed. In a first phase IoT frameworks are offer as a full stack implementation for ICOs (from collection, processing and deployment up to service and application delivery). The next step is to design and deploy horizontal IoT frame works that support on-demand access to the deployed IoT framework(s) (also called IoT silos), and finally the cloud infrastructures must interact each other in order to share data and management operations. Cloud-based IoT services not only can be deployed over multiple infrastructure providers (such as smart cities, municipalities and private enterprises) but over different technology (mobile, wireless, Internet). This chapter discusses research advances towards the formulation of federated cloud services support for the Internet of Things. This chapter presents the architectural components of a federated framework, which can be used to emphasize on-demand establishment of IoT cloud-based services. The service requirements from analyzing the state-of-the-art and efforts towards the convergence of cloud computing and IoT are introduced and discussed in the framework of the awarded open source rookie of the year OpenIoT project. By reviewing cloud design principles for enterprise environments and best management practices for converging utility-driven IoT infrastructures this chapter aims to provide insights on interconnection of IoT silos by means of federated cloud service management.

Keywords: Utility-Driven, Cloud Services, Internet of Things, Future Internet, Enterprise systems, Linked Data, Cloud Computing, Wireless Sensor Networks, Open Source.

9.1 Introduction

In the same way Internet services rely on communication networks, the deployment of Internet of Things (IoT) services and applications demands reliable communications infrastructure [1]. Every day cloud systems are more adopted as part of the infrastructure support and architecture design for IoT. Cloud systems play a crucial role in the Future Internet (FI) and particularly as result of the current accelerated race for providing utility-driven integrated solutions wherecloud-based enterprise services are just an example. IoT services (mainly over sensor networks) [2], are in their way to be deployed widely, however a full service deployment over multiple device technology has not been fully deployed yet. In this evolution process, it is envisioned that the deployment and provisioning of cloud-based IoT services would greatly benefit from horizontal open framework (platform) in order to deliver on-demand access to Internet of Things (IoT) devices. Cloud-based services not only can be deployed over multiple infrastructure providers (such as smart cities, municipalities and private enterprises) but different technology (3G mobile, WiFi wireless, WiMax Internet and ULE etc.).

Despite the proliferation of cloud computing models and infrastructures [3], there is still no easy way to formulate and manage IoT based cloud environments i.e. environments comprising IoT "entities" and resources (such as sensors, actuators and smart devices) and dynamically offering on-demand utility-based (i.e. pay-as-you-go) services. Up-to-date several researchers have described the benefits of a pervasive (sensor-based) distributed computing infrastructure [4–6] without however providing a systematic and structured solution to the formulation and management of utility-based IoT environments. Similarly, recent state-of-the-art participatory sensing infrastructures and services [7–9], provide instantiations of cloud-based and utility-based sensing services (e.g. Location-as-a-Service) [10], without however providing any middleware framework and disciplined approach to deploying and providing such services. Despite the rising popularity of the IoT and more recently cloud services, the concept of utility-driven cloud services is still immature and influenced by different definitions and implementations, like sensors networks [11–12], IoT platforms [13–14], and social networks of objects [15]. However, these frameworks do not provide the tools or mechanism for orchestration capabilities, in response to end-user requests for IoT services. Furthermore, they lack interoperability functionalities, which could drive a number of integrated and federated added-value features.

This chapter is organized as follow: Section 2 introduces research advances towards the formulation of federated cloud services support for the Internet of Things. Section 3 presents the federated autonomic reference model service life cycle and its adaptations in the framework of the OpenIoT project. Section 4 introduces a organizational view that enables the IoT service lifecycle to be explicitly modelled and semantically managed following autonomic principles. Section 5 presents the cloud management architecture following the design principles for cloud services control loop. Section 6 describes the designed architecture for enabling the services about Internet of Things data in the cloud. Finally Section 7 presents the conclusions.

9.2 Federated Cloud Services Management

Particular interest for cloud computing services and the use of virtual infrastructures supporting such services is also result of the business model cloud computing offers, where bigger revenue and more efficient exploitation is envisaged [18]. Likewise particular interest exists from the industry sector (where most of the implementations are taking place) for developing more management tools and solutions in the cloud. In the other hand academic communities point towards finding solutions for more powerful computing processing and at the same time more efficient and also for a more extended interoperability concept between the cloud-based IoT platforms. Thus generally problems on manageability, control of cloud and other research challenges are being investigated.

It is anticipated that cloud computing should reduce cost and time of computing and processing [21]. However while cost benefit is reflected mainly to the end-user, from a cloud service provider perspective, cloud computing is more than a simple arrangement of mostly virtual servers, offering the potential of tailored service offerings and theoretically infinite expansion. It is a potentially large number of tailored resources, which are interacting to facilitate the deployment, adaptation, and support of services, and this situation represent significant management challenges. In management terms there is a potential trend to adopt, refine, and test traditional versus federated management methods to exploit, optimize and automate [18] the management operations of cloud computing infrastructures, however this is proving difficult to implement, so designs for management by using new methodologies and paradigms are being investigated.

Cloud management is a complex task [17] as clouds must support appropriate levels of tailored service performance to large groups of diverse users. A sector of services, named private clouds, coexists with and is provisioned through a bigger public cloud, where the services associated to those private clouds are accessed through (virtualized) wide area networks. In this section challenges for managing cloud service infrastructure are discussed in the form of scenarios, special focuses on data management systems which are essential for the provisioning and control access of virtual infrastructure resources [20], and where such systems must be able to address fundamental issues related to scalability and reliability which are inherent when integrating diverse cloud-based IoT systems.

9.2.1 Cloud Data Management

The need to control multiple computers running complex applications and likewise the interaction of multiple service providers supporting a common data centre service exacerbates the challenge of finding management alternatives for orchestrating between the different cloud-based systems and data services [33]. Even though having full control of the management operations when a service is being executed is necessary, distributing this decision control is still an open issue. In cloud management systems, supporting such complex management operations [18][20] must be addressesfocusing on the challenging problem of coordinating multiple running data applications, management operations, and while prioritizing tasks for service interoperability between systems.

An emerging alternative to solve cloud computing control, from a management perspective, is the use of formal languages as a tool for information exchange between the diverse data and information systems participating in cloud service provisioning. These formal languages rely on an inference plane [34–35] for example. By using semantic decision support and enriched monitoring information management, decision support is enabled and facilitated. As a result of using semantics a more complete control of service management operations can be offered. This semantically-enabled decision support gives better control in the management of resources, devices, systems and services, thereby promoting the management of cloud with formal models [36].

It is a need to manage the cloud in a simplified way using the mechanism to represent and contain Description Logic (DL) to operate operational rules. For example, the SWRL language [37] can be used to formalize a policy language to build up a collection of model representations with the necessary semantic richness and formalisms to represent and integrate the heterogeneous information present in cloud management operations. This approach relies on the fact that high level infrastructure representations do not use resources when they are not being required to support or deploy services [38–40]. Thus with high-level instructions the cloud infrastructure can be managed in a more dynamic way.

9.2.2 Cloud Data Monitoring

Monitoring in federated cloud is essential for automatic or autonomous adaptation to current data load, as well as to provide feedback on service logic. Scalability and security are essential for cloud monitoring. In federated systems the monitoring is done locally and the information shared are the monitored reports. Unless monitoring is a external task to the system no access to the monitored resources from outside the federated resources are available. Without solving the problems of scalability and security, tools and technologies like federated monitoring are almost impossible to be deployed in a cloud environment. Cloud monitoring requires application-level information monitored in addition to the system usage data current tools can provide.

There are several related works in this area. Lattice is a distributed monitoring framework, which was exercised and validated in computing clouds [22] and in network clouds [23]. DSMon [20] introduces system monitoring for distributed environments and mainly focuses on fault tolerant aspects; NWS [24] also provides a distributed framework for monitoring and has the ability of forecasting performance changes; When compared to DSMon and NWS, another system resources monitoring tool called DRmonitoring [16], requires less resources to run and supports multiple platforms (Linux and Windows). HP Open View [25] and IBM Tivoli [26] have been developed to ease system monitoring and are primarily targeting the enterprise application environment. Although, the commercial products are relatively portable across different operating systems, they are usually highly integrated with vendor specific applications. In the cloud environment, heterogeneity is one of fundamental requirements for monitoring tools. Therefore, the industrial tools are unsuitable for more general purpose monitoring of the cloud. GoogleApp engine [27] and Hyperic [28] both provide monitoring tools for system status such as CPU, memory, and processes resource allocations. Such system usage data can be useful for general purpose cloud monitoring, but they may not be sufficient enough for an application level manager to make appropriate decisions.

9.2.3 Cloud Data Exchange

Federation in the cloud [19] would imply a requirement where user's applications or services shall still be able to execute across a federation of resources stemming from different cloud providers. It also refers to the ability for different cloud providers to scale their service offerings and to share capabilities to combine efforts and provide a better quality of service for their customers. While the technological aspects required supporting cloud federation is an ongoing research domain, there has been little work to support the holistic end-to-end monitoring and management of federated cloud services and resources. This approach requires users and multiple providers to both delegate, share and consume each other resources in a peer-to-peer manner in a secure, managed, monitored and auditable fashion, with a particular focus on interoperability between management and resource description approaches.

Federation presents as an approach for supporting the increasingly important requirement to orchestrate multiple vendors, operators and end user interactions [29–30], and now we see the applicability of this concept in the cloud computing area. Cloud computing offers an end-user perspective where the use of one or any other infrastructure is transparent, in the best case the infrastructure is ignored by the cloud user [31]. However from the cloud operator perspective, there are heterogeneous shared network devices as part of diverse infrastructures that must be self-coordinated for offering distributed management or alternatively centrally managed in order to provide the services for which they have been configured. Furthermore, there must be support to facilitate composition of new services, which requires a total overview of available resources [32]. In such a federated system, the number of conflicts or problems that may arise when using diverse information referring to the same service or individuals with the objective of providing an end-to-end service across federated resources must be analysed by methodologies that can detect conflicts. In this sense semantic annotation and semantic interoperability tools appears as tentative approach solution and under investigation.

9.2.4 Infrastructure Configuration and re-Configuration

We can identify several cloud usage patterns based on bandwidth, storage, and server instances over time [41]. Constant usage over time is typical for internal applications with small variations in usage. Cyclic internal loads are typical for batch and data processing of internal data. Highly predictable cyclic external loads are characteristic of web servers such as news, sports, whereas spiked external loads are seen on web pages with suddenly popular content.

The cloud paradigm enables applications to scale-up and scale-down on demand, and to more easily adapt to the usage patterns as outlined above. Depending on a number or type of requests, the application can change its configuration to satisfy given service criteria and at the same time optimize resource utilization and reduce the costs. Similarly clients - which can run on a cloud as well - can reconfigure themselves based on application availability and service levels required.

9.3 Federated Management Service Life Cycle

Management and configuration of large-scale and highly distributed and dynamic applications is everyday increasingly in complexity at the network and enterprise application levels. In the current Internet typical large enterprise systems contain thousands of physically distributed software components that communicate across different networks to satisfy end-to-end services client requests. Given the possibility of multiple network connection points for the components cooperating to serve a request (e.g., the components may be deployed in different data centres), and the diversity on service demand and network operating conditions, it is very difficult avoid conflicts between different monitoring and management systems to provide effective end-to-end applications managing the network.

As depicted in Figure 9.1, the federated autonomic reference model service life cycle is depicted, In the diagram is expposed how the definition and contractual agreements between different enterprises (*1.Definition*) establish the process for monitoring (*2.Observation*) and also identify particular management dataat application, service, middleware and hardware levels (*3.Analysis*) that can later be gathered, processed, aggregated and correlated

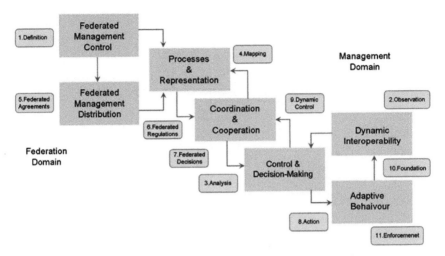

Figure 9.1   Federated Management Reference Model Life Cycle

(*4.Mapping*) to provide knowledge that supports management operations of large enterprise applications (*5.Federated Agreements*) and thus the network services that they require (*6.Federated Regulations*).

Monitoring data at the network and application level can be used to generate knowledge that can be used to support enterprise application management in a form of control loops in the information; a feature necessary in the Future Internet service provisioning process (*7.Federated Decisions*). Thus infrastructure can be re-configurable and adaptive to business goals based on information changes (*8.Action*).

It is also important to consider appropriate ways on how information from enterprise applications and from management systems can be provided to federate management systems allowing to more robustly and efficiently be processed to generate adaptive changes in the infrastructure (*9.Dynamic Control*). Appropriate means of normalising, interpreting, sharing and visualising this information as knowledge (*10.Foundations*) thus allocate new federated network services (*11.Enforcement*).

More specifically, work is being carried out to develop monitoring techniques that can be applied to record, analyse, correlate and visualise information and trends in both network management systems and enterprise application management systems, in a manner such that a coherent view of the communication profiles of different application-level and network-level services can be built.

9.3.1 Open IoT Autonomic Data Management

The first step towards federation is the automate process by integrating sensor data and cloud infrastructures. IBM[1] has introduced autonomic computing as part of the vision for "systems managing themselves according to an administrator's goals. New components integrate as effortlessly as a new cell establishes itself in the human body. These ideas are not science fiction, but elements of the grand challenge to create self-managing computing systems". This principle has emerged and transcended beyond computing frontiers and also in the area of the communications management, the term autonomic communications has been researched for several years, reflecting a real challenge to materialize the vision of transparent interaction between administrator's goals and systems self-management operations. In the late 90's supported by the Autonomic Computing Forum (ACF) autonomics brought the concept of seamless mobility associated to scenarios for people configuring new personalized services using displays, smart posters and other end-user interaction facilities, as well as their own personal devices. Named lately as pervasive computing, autonomics bring the inherent necessity to increase the functionality of those systems dealing with additional information and funded on communication system infrastructures. Pervasive service requirements are headed by the interoperability of data, voice, and multimedia using the same (converged) network. This requirement defines a new challenge: the necessity to integrate smartness to the systems and make the infrastructure more reactive by means of data and services control. Nowadays the Future Internet design with the inclusion of IoT is motivated by both, the necessity to support the requirements of pervasive services and the necessity to satisfy the challenges of self-operations dictated by the largely named Internet-of-Things paradigm.

As part of the OpenIoT and inorder to establish it as a blueprint open source solution, the creation of new challenges in terms of enhancing complex systems functionality, enable large support of sensors, devices and services systems and enable dynamic deployment and implementation is fundamental.

Autonomic systems must dynamically adapt the services and resources that they provide to meet the changing needs of users and/or in response to changing environmental conditions alike that system control demands; this requires the integration of management information. Figure 9.2 depicts the OpenIoT autonomic control loop mapping proposed in OpenIoT. This model for OpenIoT is crucial, as each day, more complex IoT consumers require services, which in turn requires more complex support systems that must

---

[1] IBM The Vision of Autonomic Computing, IBM Research, Vision and Manifesto.

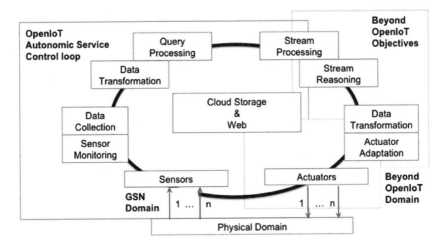

Figure 9.2 Open IoT Autonomic Service Control Loop for IoT

harmonize multiple technologies and linked information from sub-systems interacting to offer embedded services.

IoTenvisage new smart scenarios, and at the same time challenging the user-centric applications and services. IoT systems require information and systems able to support services and especially interoperable applications. In autonomic systems linked data plays the important role of enabling the management plane to adapt the services and resources that it is offering to the changing demands of the user, as well as adapt to changing environmental conditions, by meaning of the linked nature, thus enabling the management of new functionalities in IoT complex system [45].

9.3.2 Performance

Scalability and Interoperability between heterogeneous, complex and distributed Internet-connected objects systems is always a challenge and it requires new management and optimization functionalities. As an inherent functional limitation, IoT management systems do not support a large spectrum of devices, such as wearable computers and specialized sensors. Furthermore, IoT systems are every day being provided with embedded technology/connectivity, which is used to make new types of networks that provide their own services (e.g., simple services supporting other, more complex, services), which this converts the management task more difficult and complex in terms of scalability.

In OpenIoT, we deal with linked data or information sharing. In the project scenarios we are typified a broad mixture of technologies and devices (sensors) that generate an extensive amount of different types of information, many of which need to be shared and reused among the different service management components with different data representation-sharing mechanisms. This requires the use of different data models, due both to the nature of the information being managed as well as the physical and logical requirements of applications. However, information/data models (linked data and particularly RDF) do not have everything necessary to build up this single common interoperable sharing support system. In particular, there is a need of delegate the ability to describe behaviour of the services and application with the infrastructure.

9.3.3 Reliability

Traditionally management systems approaches define a strict layering of functionality and cross-layered interactions left beside, in OpenIoT the broad diversity of resources, devices, services, and infrastructure systems interconnected and exchanging information is used.

In OpenIoT the objective of annotating information, described in services and data models, can provide an extensible, reusable common manageability platform that provides new functionality to better manage resources, devices, networks, systems and services [45]. Given the fact that different data representations are a necessity in the next generation Internet solutions (Clark 2003), the typical solutions have attempted to define a single common information model that can harmonize the information present in each of these different management data models. Using a single information model prevents different data models from defining the same concept in conflicting ways. In addition, the use of a single common information model enables the reuse and exchange of service management information. Examples of using a single common information model include the initiative CIM/WBEM (Common Information Model/Web Based Enterprise Management), (DMTF-CIM) from the DMTF (Distributed Management Task Force, Inc) and broadly supported by the Shared Information Model (TMF-SID) of the TMF (Tele Management Forum). However neither has been completely successful, as evidenced by the lack of support for either of these approaches in network devices currently manufactured. This indicates that SID model lacks the extensibility to promote the interoperability and enhance its acceptance and expand its standardization.

Other examples in the level of software operation and applications with other technologies are Microsoft DCOM[2] (Distributed Component Object Model - Microsoft) or Java RMI[3] (Remote Method Invocation - Sun Micro Systems) that are being used in many applications. Even those initiatives do not allow sharing the information with each other technologies freely.

In OpenIoTis proposed as alternative to facilitate the interoperability, the use of linked data for semantically enriching the information models to contain the references in the form of relationships between the necessary sensor data required to provide the service. By using one or more ontologies and the referenced sensor data ontology (W3C SSN)[4] then services systems and applications using information contained in the service model can access and do operations and functions for which they were designed. This functionality is in particular impacting the performance of the ICO systems as per it unique and novelty feature of enabling management operations using the information contained in the information models (sensor data) for ICO service provisioning.

9.3.4 Scalability

The vision of the IoT, which enables societies to use a wide range of sensors, devices and computing systems to "transparently" create smart applications and on-demand services automatically, requires beyond sub-systems offering reliable control and connectivity, associated management systems that are able to support such exponentially growing and dynamic services creation. However, the multiplicity and heterogeneity of technologies used, such as wireless, fixed networks and mobile devices that can use both, is a barrier to achieving seamless interoperability.

In OpenIoT information data modelling is formalised by means of ontologies (SSN and others), with policy information from information and data models, to manage IoT applications and services to create more extensible information models based on linked data solutions. This approach uses information to represent, describe, contain share and reuse information. This OpenIoT approach creates the basis to find the best way to integrate information into service operations.

---

[2]http://www.microsoft.com/com/
[3]http://www.oracle.com/technetwork/java/javase/tech/index-jsp-136424.html
[4]http://www.w3.org/2005/Incubator/ssn/

The Internet affects management of IoT devices as well as their services platforms and its new business models demands as web services (W3C-WebServices)[5] As result, new descriptions of operations and management functions for supporting IoT Devices are necessaries, also the organization of data seeking information interoperability. These aspects affect both the organizational view of the service lifecycle as well as its operational behaviour, and use semantic control in each to achieve interoperability in the information necessary to control and manage pervasive services.

9.3.5 Resource Optimization and Cost Efficiency

Self-management features depends on both the requirements as well as the capabilities of the middleware frameworks or platforms for managing information describing the services as well as information supporting the delivery and maintenance of the services. The representation of information impacts the design of novel syntax and semantic tools for achieving the interoperability necessary when ICO resources and services are being managed. Middleware capabilities influence the performance of the information systems, their impact on the design of new services, and the adaptation of existing applications to represent and disseminate the information.

In OpenIoT, the use of rules-based engines for controlling IoT service management is augmented with the use of standard ontologies. This enables the management systems to support the same management data to accommodate the needs of different management applications through the use of rich semantics [18]. Service management applications for IoT systems highlight the importance of formal information.

The rules are used in the managing of various aspects of the lifecycle of services. It is important to identify in OpenIoT what is meant by the term "service lifecycle". Currently, the TMF is specifying many of the management operations in networks for supporting services (TMN-M3050)(TMN-M3060), in a manner similar to how the W3C specifies web-services (W3C-WebServices). However, a growing trend is to manage the convergence between infrastructure and services (i.e., the ability to manage the different service requirements of data, voice, and multimedia serviced by the same network), as well as the resulting converged services themselves. The management of NGN pervasive services involves self-management capabilities for improving performance and achieving the interoperability necessary to support current and next generation services.

---

[5]http://www.w3.org/2002/ws/.

9.4 Self-management Lifecycle

This section describes an organizational view that enables the IoT service lifecycle to be explicitly modelled and semantically managed. This in turn ensures information interoperability necessary to manage different services in IoT applications. This section describes the organizational view for the Autonomic Self-management Framework, which can be divided into six distinct phases with specific tasks [45].

Management operations enabling the autonomic nature of IoT systems are the core part of the IoT service lifecycle, and where the contributions in OpenIoT are focused. Thus, the management phase is highlighted in Figure 9.3 regarding creation and customization of services, accounting, billing and customer support are outside the scope of OpenIoT however considered from a design description for IoT systems. The different service phases exposed in this section describe the service lifecycle foundations. The objective is focusing the research efforts in understanding the underlying complexity of service management, as well as to better understanding about the roles for the components that make up the service lifecycle, using interoperable information that is independent of any specific type of infrastructure that is used in the deployment of IoT services.

9.4.1 Service Creation

The creation of each new IoT service starts with a set of requirements; the service at that time exists only as an idea. This idea of the service originates from the requirements produced by market analysis and other business

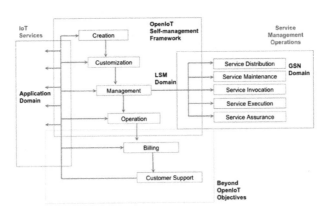

Figure 9.3 Open IoT Autonomic Self-management Framework forIoT Services (ICO's)

information. At this time, technology-specific resources are not considered in the creation of a service. However, the infrastructure for provisioning this service must be abstracted in order to implement the business-facing aspects of the service as specified in a service definition process [45].

The idea of IoT service must be translated into a technical description of a new service, encompassing all the necessary functionality for fulfilling the requirements of that service (e.g. physical devices interconnection, sensor data collection, virtual sensor aggregation, etc.). A service is conceptualized as the instructions or set of instructions to provide the necessary mechanism to provide the service itself and called service logic (SLO).

9.4.2 Efficient Scheduling

The OpenIoT system comprises the notion of scheduling of requests which undertakes the task of technically describing a new service. The OpenIoT global scheduler component, which OpenIoT architecture specifies, receives all the User requests for IoT services and fulfils the requirements of that service. A wide range of different optimization algorithms can be implemented at the scheduler component of the OpenIoT architecture. Moreover, OpenIoT defines a local scheduler at the level of its sensor middleware node (Virtual Sensors). This node has control over the sensors therefore appropriate local-level optimizations can be implemented. So the main OpenIoT efficient multi-level (global, local) scheduling optimization scheme involve multi-query data management and caching techniques.

9.4.3 Service Customization

Service customization, which is also called authoring, is necessary for enabling the IoT service provider to offer for its consumers the ability to customize aspects of their IoT services (i.e. ICO selection and/or configuration) according to their personal needs and/or desires (e.g. defined by a query language). Today, this is a growing trend in web-services and business orientation. An inherent portion of the customization phase is an extensible infrastructure, which must be able to handle service subscription and customization requests from administrators as well as ICO consumers.

9.4.4 Efficient Sensor Data Collection

In OpenIoT we focus on stream data processing components enabling the deployment over multiple infrastructures. By combining query languages (i.e. SPARQL) and stream data processing components the end user can customize services on its own needs. Based on an efficient stream proceeding, at the collection and distribution, the sensor data system are more efficient towards previous identification of "intelligent" data providers' by sensing and actuating over streaming data infrastructures, rather than having to deploy data processing infrastructures by themselves.

9.4.5 Request Types Optimization

Another type of service customization in OpenIoT exists in the request types optimization where we make use of LSM [46]. We use LSM as an extended middleware with functionalities to transparently cater for dynamic stream information. LSM is using efficient query algorithms that may provide to the data processing operators a global view of the whole dataset.

9.4.6 Service Management

In this section, the management operations of an ICO service and its inter-actions are identified as distinct management operations from the rest of the service lifecycle phases. Figure 9.4 depicts management operations as part of the management phase in a pervasive service lifecycle.

The main service management tasks are service distribution, service maintenance, service invocation, service execution and service assurance. An important functional aspect of the OpenIoT service management framework implementation is the dynamic on the fly deployment of IoT services using specific logic rules. For instance, when an IoT service is going to be deployed, decisions have to be taken in order to determine which sensor or devices

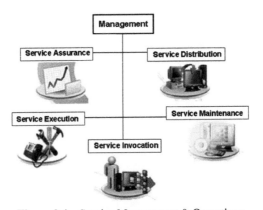

Figure 9.4   Service Management & Operations

(things) are going to be used to support the service. This activity is most effectively done through the use of particular logic rules that map the user with the desired data sources and with the capabilities of the set of ICO that are going to support the service. Moreover, service invocation and execution can also be controlled by same logic rules, which enable a flexible approach for customizing one or more service templates to multiple users.

In the other hand, management is an effective mechanism for maintaining code to realize the IoT services, changes and assurance of the IoT service included. For example, when variations in the delivery of the service are sensed by the system, one or more policies can define the set of actions that need to be taken to solve the problem. In this way, the use of policies enables different behaviour to be orchestrated as a first step to implement self-management functionality.

9.4.6.1 Service Distribution

This step takes place immediately after the service creation and customization in the service lifecycle. It consists of storing the service code in specific storage points. Policies controlling this phase are termed code distribution policies (Distribution). The mechanism controlling the code distribution determines the specific set of storage points that the code should be stored in. The enforcement is carried out by the components that are typically called Code Distribution Action Consumers.

9.4.6.2 Service Maintenance

Once the code is distributed, it must be maintained in order to support updates and new versions. For this task, we use special policies, termed code maintenance Policies (CMaintenance). These policies control the maintenance activities carried out by the system on the code of specific services. A typical trigger for these policies could be the creation of a new code version or the usage of a service by the consumer. The actions include code removal, update and redistribution. These policies can then be enforced by the component that is typically named the Code Distribution Action Consumer.

9.4.6.3 Service Invocation

The service invocation is controlled by special policies that are called SInvocation Policies. The service invocation tasks are realized by components named Condition Evaluators, which detect specific triggers produced by the service consumers. These triggers also contain the necessary information that policies require in order to determine the associated actions. These actions consist of addressing a specific code repository and sending the code to specific execution environments in the network. The policy enforcement takes place in the Code Execution Controller Action Consumer.

9.4.6.4 Service Execution

Code execution policies, named CExecution policies, govern how the service code is executed. This means that the decision about where to execute the service code is based on one or more factors (e.g., using performance data monitored from different network nodes, or based on one or more context parameters, such as location or user identity). The typical components with the capability to execute these activities are commonly named Service Assurance Action Consumers, which evaluate network conditions. Enforcement of these policies are embedded with the responsibility of the components that are typically called Code Execution Controller Action Consumers.

9.4.6.5 Service Assurance

This phase is under the control of special policies termed service assurance policies, termed SAssurance, which are intended to specify the system behaviour under service quality violations. Rule conditions are evaluated by the Service Assurance Condition Evaluator. These policies include preventive or proactive actions, which are enforced by the component typically called the Service Assurance Action Consumer. Information consistency and completeness is guaranteed by a policy-driven system, which is assumed to reside in the service creation and customization framework.

9.4.7 Utility-based Optimization

In OpenIoT for the dynamic deployment of IoT services we adapt a utilitarian approach optimization for the system's logic rules. The utilitarian approach tries to maximize the net benefit measured as difference between the benefit of the provided information and the cost of maintaining the system in terms of energy consumption/bandwidth and the cost of ensuring privacy.

9.4.7.1 Cloud Optimization

In OpenIoT we enforce adaptive cloud optimization algorithms based on the needs of each deployed scenario. The cloud infrastructure is monitored based on its functional schemes (i.e. access/storage charges) and adapts to the service runtime having in mind its cost-effectiveness and data integrity /security.

9.4.8 Service Operation

The operation of a deployed IoT service is based on monitoring aspects of the cloud infrastructure that support that service, and variables that can modify the features and/or perceived status of the communications. Usually, monitoring tasks are done using agents, as they are extensible and can only accommodate a wide variety of information, and are easy to deploy. The information is processed by the agent and/or by middleware that can translate raw data into and from having explicit semantics that suit the needs of different applications.

9.4.8.1 Service Billing

Service billing is just as important as service management, since without the ability to bill for delivered IoT services provided, the organization providing those services cannot make money. Service billing is often based on using one or more accounting mechanisms that charge the customer based on the resources used in the network. In OpenIoT, we particularly align our approach with the cloud paradigm enabling pay-as-you-go services. In the billing phase, the information required varies during the business lifecycle, and may require additional resources to support the billing. Service metering is an module that is part of the utility manager, which keeps track of the utility metrics specified on each application. This metering can then serve as a foundation for implementing the service billing module(s).

9.4.9 Customer Support

Customer support provides assistance with purchased IoT services, while IoT main feature is the non-dependence or dependency of service provider, computational[6] resources or software[7] http://en.wikipedia.org/wiki/Software, or other support goods are required for the provisioning of complex IoT services. Therefore, a range of services[8] http://en.wikipedia.org/wiki/Customer_service and resources (mainly cloud) related are required to facilitate the maintenance and operation of the IoT services, and additional context (and sometimes the uncovering of implicit semantics) is necessary in order for user or operators to understand problems with purchased services and resources. OpenIoT foresees to enable the User with the ability to configure, monitor and maintain

---

[6]http://en.wikipedia.org/wiki/Computer
[7]software
[8]services

IoT operative services This is done through specialized monitoring and configuration interfaces which it is capable, for example, to modify object-objet connections or activate/de-active sensors, instead of relaying this capacity to the implemented service maintenance functionality in the subsystem. This maintain theOpenIoT platform system administrator/ service provider tool, which would enable him to deploy, configure, manage and offer service customer support more dynamically when necessary.

9.5 Self-Organising Cloud Architecture

In cloud computing, highly distributed and dynamic elastic infrastructures are deployed in a distributed manner to support service applications. In consequence development of management and configuration systems over virtual infrastructures are necessary. Cloud computing typically is characterized by large enterprise systems containing multiple virtual distributed software components that communicate across different networks and satisfy particular but secure personalized services requests [23][43]. The complex nature of these users request results in numerous data flows within the service components and for this reason the cloud infrastructure cannot be readily correlated with each other.

At the Figure 9.5 the cloud management architecture following the design principles for cloud services control loop is depicted. From a data model perspective, on this control loop, on-demand scalability and scalability by computing data correlation between performance data models of individual components and service management operations control are addressed. Exact component's performance modeling is difficult to achieve since it depends on diverse number of variables that range from used technology to other technology. To simplify this complexity the cloud service lifecycle model rather focuses on standard *Performance Analysis* such as available memory, CPU usage, system bus speed, and memory cache thresholds. Instead of exact performance it is also most practical to use an estimated model calculated based on monitored data from the *Data Correlation Engine* represented in the Figure 9.5 [42].

Management operations modifying the cloud service lifecycle control loop and satisfying user demands, about quality of service and reliability, play a critical role in the design of the management systems by the inherent complexity associated with the management processes of the infrastructure itself and the *Cloud Service Lifecycle Operations and Control*.

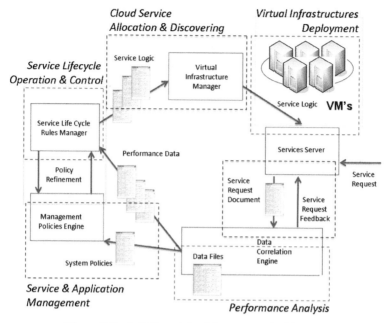

Figure 9.5 Self-Organizing Management Architecture

9.6 Horizontal Platform

OpenIoT can be seen as a framework for the convergence of cloud computing and the Internet of Things (IoT). This convergence needs to harmonize the radical differences and conflicting properties of pervasive technology (e.g., sensors and WSN (Wireless Sensor Networks) and virtual environments (Private and Public Cloud environments). Indeed, sensor networks are location-dependent, resource-constrained and expensive to develop and deploy. On the other hand, cloud-computing infrastructures are location-independent, elastic and provide access to a multitude of computing resources. Additionally, the sensor data and the information used at the application level for providing IoT services most of the times is incompatible.

OpenIoT bridges the associated gaps within these differences, since it allows IoT sensor data solutions (i.e. the Global Sensor Network (GSN) middleware solution) to leverage the sensor data to the rapid elasticity of data storage and processing in clouds in order to store the abundance of sensor data streams that are produced in the scope of large-scale deployments. Moreover, computing resources of a cloud could be also used to facilitate stream processing and management (especially in the case of computationally intensive signal processing algorithms). Overall, OpenIoT connects sensors with cloud computing infrastructures, while at the same time providing service-based access to sensor data and resources (notably based on the REST (REpresentation State Transfer) protocols.

OpenIoT is a designed architecture for enabling the services about Internet-of-Things data in the cloud, the middleware development and specification efforts have undertaken for 24 months resulting in a number of integrated software modules, components and interfaces for architecting and integrating IoT cloud-based data applications. The efforts for integrating the software components, which is however limited to providing the means for instantiating a specific IoT architecture that is appropriate for a set of concrete use case problems at hand, without a focus on the interoperability of legacy architectures and standards-based solutions. In a sense, OpenIoT has adopted a service-layered top-down solution, which can lead to the development/instantiation of compatible bottom-up architectures [47].

OpenIoT enables service providers to deploy dynamically IoT sensor data services running on cloud/utility-based infrastructure through responding to appropriate end-user requests enabling the dynamic, self-organizing and self-managing of cloud environments for IoT. The OpenIoT middleware framework therefore serves as a blueprint for non-trivial IoT applications, which are delivered according to a utility (pay-as-you-go) model.
OpenIoT addresses the following key research issues:

1. The dynamic formulation of utility-based computing environments of Internet-connected objects, in response to an **Autonomic** behaviour by following dynamically defined end-users' requests.
2. IoT Sensor Data applications provided as a service (e.g., Sensing-as-a-Service) over dynamically created and configured societies of "things" and according to a **Cloud/Utility Based** model "pay-as-you-go" (especially for enterprise applications).
3. Provisioning as a royalty-free implementation, the OpenIoT approach follows **Open Source** project rules and is built and prototyped over existing popular open source middleware platforms for RFID/WSN, notably the Global Sensor Networks (GSN) and the AspireRFID platforms.
4. The dynamic orchestration of Internet-connected objects and related resources in the cloud environment. This **Dynamic** orchestration enables response to dynamically defined end-users' service requests.
5. The OpenIoT middleware platform takes into account constraints associated with energy efficiency and bandwidth resources optimization for offering **Optimal and Self-Managing** capabilities.

6. The support for IoT applications involving trillions of things, which is geographically and administratively dispersed (as part of **Scalable** inter-domain environments).
7. The inherent **Secure, trustworthy and privacy friendliness**. The project is investigating the economics of privacy and security with a view researched utility metrics of the cloud infrastructure.
8. The associated idea for numerous sensors and mobile devices jointly collecting and sharing data of interest to observe and measure phenomena over a large geographic area. **Mobility** and **Quality of Service (QoS)** for the IoT through efficient algorithms for continuous processing and filtering of sensor data streams. Algorithms based on publish/subscribe principles that are tailored to the OpenIoT cloud to support data acquisition from wearable sensors and mobile phones as well as elastic processing of sensor data. Application-level QoS and SLA issues related to the utility-based provisioning for IoT applications.

As part of the specification process of the OpenIoT architecture and the initial exploration of the basic functionalities for IoT systems defined in the scope of the FP7 IOT-A project, The OpenIoT architecture adopts the Architecture Reference Model (IOT-A's ARM). Understanding the functional blocks and defining the data flow process in the OpenIoT Architecture (Figure 9.2) helps to understand the implemented OpenIoT service delivery mechanisms. In particular, service delivery is based on the interactions and information exchange between the OpenIoT components. The interactions are detailed in following paragraphs.

An initial setup of the OpenIoT core components and with minimum configuration files (0) the IoT system gets ready. By following End User IoT services request (1), a dynamically-generated query is defined for collecting the data content (2). A Discovery process of the available sensors (Data Services) is instantiated (3). If there are available sensors the data is collected otherwise a Sensor Configuration process starts (4). Once the data services are defined (sensors are available and reserved) a collection of data (content) is performed (5). When a process of content adaptation (data transformation) is necessary it is performed (6), otherwise the utility service is evaluated (7). The delivery of the service is performed (8), and the results of this operation collected, analysed and ready to be shown as results of the Service Visualization (9). Once the visualization (10) is performed the presentation of data is ready (11) and by means of specific utility metrics the service quality is reported (12), allowing, in this form evaluate the service, and the quality of data in order to keep records for future service provisioning requests.

In the following section, an overview of the main components of the OpenIoT architecture is provided.

9.6.1 Open IoT Architecture: Explanation and Usage

OpenIoT project identified requirements from the various IoT stakeholders (i.e. end user, service providers and programmers), to produce the main technical specifications of the OpenIoT platform and specify the functional and non-functional requirements of the OpenIoT validating applications. As part of the results in the projects is the full set of specification for a modular architecture that drives the integration of the project in terms of the implementation of both the OpenIoT platform (primarily) and of the OpenIoT applications (secondarily). As part of the OpenIoT architecture the project identified and detailed the functionalities of the main modules of the OpenIoT platform that are described in the public deliverables for the OpenIoT architecture specification. Figure 9.6 presents these components and their structuring in the scope of the OpenIoT architecture.

OpenIoT have designed and implemented ontologies for representing sensors and sensor data for storage and exchange between the different core components of the OpenIoT architecture. The OpenIoT ontology is based on ontologies for sensors and sensor data the SSN-XG ontology [48].

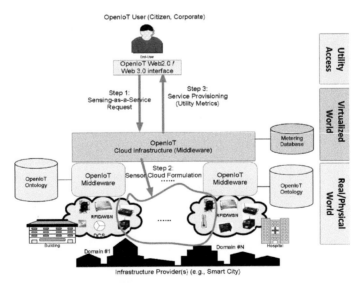

Figure 9.6   OpenIoT project – IoT Interoperability aim(s) Representation

We have extended GSN (Global Sensor Network) with RDF features and Linked Data, and provided a GSN-CoAP wrapper to access sensor data using a RESTful approach. We also investigated the use of REST interfaces and Linked Data principles to represent sensor information. This led to a first design of a URI scheme to represent sensor features on the Web, and the representation and interlinking of sensor descriptions with other Linked Open Data (LOD) datasets.

We have already implemented and created a prototype for the edge intelligence server that supports annotation of different types of sensors and enable as basic functionality finding sensors (Figure 9.7). The Edge Intelligent Server offers as an output a certain sensor value at the time of running a query over the data the sensors is "observing" based on the idea/design that sensor data is temporally and/or spatially correlated (real time streaming data). We have tested the Edge Intelligence Server prototype in private cloud infrastructure and started trials on public cloud (e.g., Amazon EC2). We have started to investigate on sharing sensor information and set up a large-scale (a.k.a. Open Public Cloud) demo for experiments. We also started the

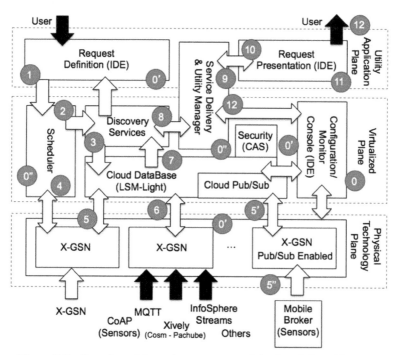

Figure 9.7 Overview of the main components of the OpenIoT Architecture

progress towards enabling the application of reasoning algorithms for seamless automated information exchange between networks of Internet-connected objects. Detailed description and components specification can be found at the OpenIoT project Deliverables available via OpenIoT project Website.

9.6.2 Cloud Services for Internet-connected objects (ICO's)

A prototype have been designed and implemented as an infrastructure for scheduling requests for IoT services (i.e., the OpenIoT scheduler) (Figure 9.8). This infrastructure enables the processing of requests for IoT services, the dynamic discovery of the sensors and resources needed to realize a services, as well as the reservation of the resource entailed in the service.

We designed and prototyped implementation a module enabling the execution and delivery of IoT services (e.g., execution of SPARQL queries), as well as the maintenance of information for utility metrics calculation on the basis of an appropriate middleware infrastructure (Figure 9.8).

We designed and prototyped a range of visual tools enabling end users' and developers' interactions with IoT services (Figure 9.8). These include tools for defining/specifying IoT services (e.g., the request definition tool), tools for defining and announcing sensors (e.g., the schema editor tool), as well as tools for the management of the OpenIoT middleware infrastructure.

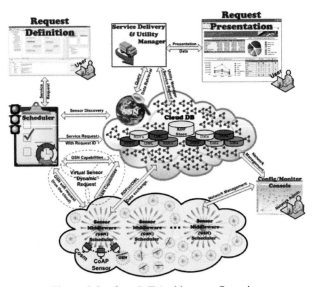

Figure 9.8 OpenIoT Architecture Overview

We designed and implemented a library of visualization components (Figure 9.8), which facilitate the presentation of IoT services that are developed, deployed and enacted over the OpenIoT infrastructure.

We worked on the full integration of the OpenIoT middleware platform on the basis of components and modules developed in WP4 (i.e., scheduler, GUIs, utility manager) and the technical results of the project that have been developed in WP3 (i.e., OpenIoT ontology, edge intelligence (LSM–Light middleware, extended GSN middleware (X-GSN)).

9.6.3 Management of IoT Service Infrastructures following Horizontal Approach

We have designed a framework for managing the OpenIoT service infrastructure, focusing on two main aspects: a self-management and optimization framework (Figure 9.6), and a privacy and security framework (Figure 9.9).

We have investigated different techniques for optimizing resource sharing on the OpenIoT platform. More specifically, we have proposed specific algorithms based on utility functions for maximizing the overall social

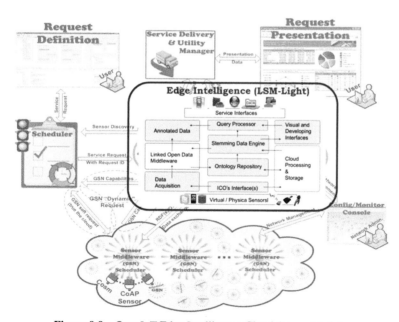

Figure 9.9 OpenIoT Edge Intelligence Cloud-Server Module

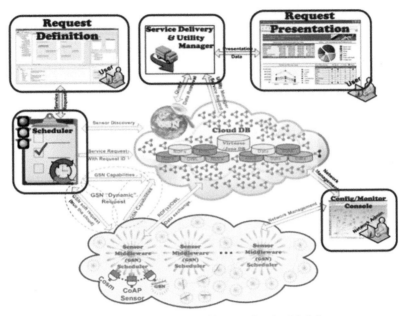

Figure 9.10   OpenIoT Architecture Service Modules

welfare, taking into account different types of queries posed by the users, as well as cost and valuation functions.

We have also proposed cloud optimization storage algorithms and techniques for model-view sensor data. The core of this approach is to fragment sensor time series and then approximate each data segment by a mathematical function with certain parameters, such that a specific error norm is satisfied. Then we have designed and prototyped a memory and index cloud-storage framework that is able to answer queries using the segment approximations.

We have investigated query cache mechanisms for IoT queries in the cloud, and the trade-offs of using cache compared to a cloud data store. To demonstrate and simulate such scenarios, a simulator has been created that models the various costs, assessing the overall cost efficiency of the system when using a cache.

We have we proposed a central authorization and authentication server that provides authentication and authorization services for all relevant OpenIoT applications running on behalf of different subjects. The main feature of this architecture is that user credentials (username/password) are only checked and maintained by a Central Authorization Server (CAS) and it authorizes applications running on behalf a user by granting them an access token with a

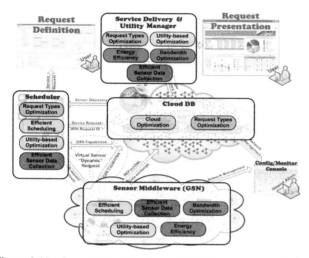

Figure 9.11 Open IoT Architecture with Self-management Features

given time to live. This prevents any circulation of the credentials throughout OpenIoT components.

The vision of self-management creates an environment that hides the underlying complexity of the management operations, and instead provides a façade that is appealing to both administrators and end-users alike. It is based on consensual agreements between different systems (e.g., management systems and information support systems), and it requires a certain degree of cooperation between the systems to enable interoperable data exchange.

One of the most important benefits of this agreement is the resulting improvement of the management tasks and operations using such information to control ICO's and their applications. However, the descriptions and rules that coordinate the control operations of an ICO system are not the same as those that govern the sensor data in each application system. For example, information present in a particular sensor network with primarily proprietary technology is almost restricted for using to control the operation of a service, and usually has nothing to do with managing the service.

In the scope of OpenIoT, cooperation and interactions between the various components of the OpenIoT architecture are required in order to support the self-management functionalities.

9.7 Conclusions and Future Work

This chapterdiscussed some of the key state of the art developments on federated cloud service management and its applicability to the Internet of Things, including concepts, trends, challenges and limitations in the area of cloud computing. One of the most important research challenges in cloud systems is the dynamic control of elasticity of the cloud infrastructure. Based on performance metrics (i.e. log files from computing applications) or data processing request (i.e. annotated data as streaming data analysis).

We have introduced and discussed advances in terms of enabling federation by using link data mechanisms. Implemented architectural components, in the form of prototyped middleware solutions have been presented and evaluated as part of proof of the concept process enabling interoperability of the cloud infrastructure. In addition by using linked data mechanisms maximising control of service lifecycle can be achieved.

It has been discussed IoT frameworks are currently looking to be offered as a full stack implementation for IoT (from collection, processing and deployment up to service and application delivery). The next step is to design and deploy horizontal IoT frameworks that support on-demand access to the deployed IoT framework(s) (also called IoT silos), and finally the cloud infrastructures must interact each other in order to share data and management operations. Cloud-based IoT services not only can be deployed over multiple infrastructure providers (such as smart cities, municipalities and private enterprises) but over different technology (mobile, wireless, Internet).

Further work includes experimentation with service-discovering mechanism by using service allocation protocols and embedded optimisation methods and algorithms for link data usage in autonomic service management.

Acknowledgments

Part of this work has been carried out in the scope of the project ICT OpenIoT Project (Open source blueprint for large scale self-organizing cloud environments for IoT applications), which is co-funded by the European Commission under seventh framework program, contract number FP7-ICT-2011–7-287305-OpenIoT.

References

[1] Harald Sundmaeker, Patrick Guillemin, Peter Friess, Sylvie Woelf-flé (eds), "Vision and Challenges for Realising the Internet of Things", March 2010, ISBN 978–92-79–15088-3, doi:10.2759/26127, ©European Union, 2010.

[2] K. Aberer, M. Hauswirth, and A. Salehi. "Infrastructure for data processing in large-scale interconnected sensor networks". In MDM'07, pages 198–205, 2007.

[3] Michael Armbrust, Armando Fox, ReanGrith, Anthony D. Joseph, Randy H. Katz, Andrew Konwinski, Gunho Lee, David A. Patterson, Ariel Rabkin, Ion Stoica, and MateiZaharia. "Above the Clouds: A Berkeley View of Cloud Computing". Technical Report UCB/EECS-2009–28, EECS Department, University of California, Berkeley, Feb 2009.

[4] Hock Beng Lim, Yong MengTeo, Protik Mukherjee, Vinh The Lam, Weng Fai Wong, Simon See, "Sensor Grid: Integration of Wireless Sensor Networks and the Grid", Proceedings of the The IEEE Conference on Local Computer Networks 30th Anniversary, p.91–99, November 15–17, 2005, doi>10.1109/LCN.2005.123.

[5] Lina Yu, Xiang Sun, Qing Wang, Wanlin Gao, Ganghong Zhang, Zili Liu, Zhen Li, Jin Wang, "Research on Resource Directory Service for Sharing Remote Sensing Data under Grid Environment", gcc, pp.344–347, 2009 Eighth International Conference on Grid and Cooperative Computing, 2009.

[6] Kapadia, A.; Myers, S.; XiaoFeng Wang; Fox, G. "Secure cloud computing with brokered trusted sensor networks", International Symposium on Collaborative Technologies and Systems (CTS), 17–21 May 2010, pp.581 – 592.

[7] Catherine Havasi, James Pustejovsky, Robert Speer, Henry Lieberman, "Digital Intuition: Applying Common Sense Using Dimensionality Reduction", IEEE Intelligent Systems 24(4), pp. 24–35, July 2009.

[8] Nicolas Maisonneuve, Matthias Stevens, Maria E. Niessen, Luc Steels, "NoiseTube: Measuring and mapping noise pollution with mobile phones", Information Technologies in Environmental Engineering (ITEE 2009), Proceedings of the 4th International ICSC Symposium, Thessaloniki, Greece, May 28–29, 2009.

[9] Page, X. and A. Kobsa "Navigating the Social Terrain with Google Latitude".iConference 2010, Urbana-Champaign, IL, p.174–178.

[10] Marshall Kirkpatrick "The Era of Location-as-Platform Has Arrived", ReadWriteWeb, January 25, 2010

[11] Mohammad Mehedi Hassan, Biao Song, Eui-nam "Huh: A framework of sensor-cloud integration opportunities and challenges". ICUIMC 2009: 618–626

[12] MadokaYuriyama, Takayuki Kushida "Sensor-Cloud Infrastructure - Physical Sensor Management with Virtualized Sensors on Cloud Computing". NBiS 2010: 1–8

[13] Parwekar, P, "From Internet of Things towards cloud of things", 2nd IEEE International Conference on Computer and Communication Technology (ICCCT), September 2011.

[14] Pachube Internet of Things "Bill of Rights" Ed Borden, in pachube.com (2011)

[15] Matthias Kranz, Luis Roalter, and Florian Michahelles, "Things That Twitter: Social Networks and the Internet of Things", What can the Internet of Things do for the Citizen (CIoT) Workshop at The Eighth International Conference on Pervasive Computing (Pervasive 2010), Helsinki, Finland, May 2010.

[16] The 'Intercloud' and the Future of Computing, an interview: Vint Cerf at FORA.tv, the Churchill Club, Jan 7, 2010. SRI International Building, Menlo Park, CA, Online Jan 2011. http://www.youtube.com/user/ForaTv#p/search/1/r2G94ImcUuY

[17] B. Rochwerger, J. Caceres, R.S. Montero, D. Breitgand, E. Elmroth, A. Galis, E. Levy, I.M. Llorente, K. Nagin, Y. Wolfsthal, E. Elmroth, J. Caceres, M. Ben-Yehuda, W. Emmerich, F. Galan. "The RESERVOIR Model and Architecture for Open Federated Cloud Computing", IBM Journal of Research and Development, Vol. 53, No. 4. 2009.

[18] Serrano, J. Martin "Applied Ontology Engineering in Cloud Services, Networks and Management Systems", to be released on March 2012, Springer Publishers, 2012. Hardcover, p.p.222 pages, ISBN-10: 1461422353, ISBN-13: 978–1461422358.

[19] Chapman, C., Emmerich, E., Marquez, F. G., Clayman, S., Galis, A. - "Software Architecture Definition for On-demand Cloud Provisioning" - Springer Journal on Cluster Computing – DOI: 10.1007/s10586–011–0152–0; May 2011

[20] S. Clayman, A. Galis, L. Mamatas, "Monitoring Virtual Networks", 12th IEEE/IFIP Network Operations and Management Symposium (NOMS

2010) - International on Management of the Future Internet, 19–23 April'10, Osaka http://www.man.org/2010/

[21] IBM Software Group, U.S.A. "Breaking through the haze: understanding and leveraging cloud computing" Route 100, Somers, NY. IBB0302-USEN-00. 2008

[22] S. Clayman, A. Galis, G. Toffetti, L. M. Vaquero, B. Rochwerger, P. Massonet "Towards A Service-Based Internet" Lecture Notes in Computer Science, 2010, Volume 6481/2010, 215–217, DOI: 10.1007/978-3-642-17694-4_30

[23] J. Shao, H. Wei, Q. Wang, and H. Mei. A runtime model based monitoring approach for cloud. In Cloud Computing (CLOUD), 2010 IEEE 3rd International Conference on, pages 313 –320, July 2010.

[24] R. Wolski, N. T. Spring, and J. Hayes. The network weather service: a distributed resource performance forecasting service for metacomputing. Future Gener. Comput. Syst., 15:757–768, October 1999.

[25] HP. Hpopenview event correlation services, Nov. 2010. Available [online]: http://www.managementsoftware.hp.com/products /ecs/ds/ecs ds.pdf

[26] IBM. Tivoli support information center, Nov. 2010. Available [online]: http://publib. boulder.ibm.com/tividd/td/IBMTivoliMonitoring forTransactionPerformance5.3.html.

[27] Google. Google app engine system status, Nov. 2010. Available [online]: http://code.google.com/status/appengine.

[28] Hyperic. Cloudstatus ® powered by hyperic, Nov. 2010. Available [online]: http://www.cloudstatus.com.

[29] Bakker, J.H.L. Pattenier, F.J. "The layer network federation reference point-definition and implementation" Bell Labs. Innovation, Lucent Technol., Huizen, in TINA Conf Proc. 1999.p.p. 125–127, Oahu, HI, USA, ISBN: 0–7803-5785-X

[30] Serrano J.M., Van deer Meer, S., Holum, V., Murphy J., and Strassner; J. "Federation, A Matter of Autonomic Management in the Future internet". 2010 IEEE/IFIP Network Operations & Management Symposium – NOMS 2010.Osaka International Convention Center, 19–23 April 2010, Osaka, Canada.

[31] Allee, V. "The Future of Knowledge: Increasing Prosperity through Value Networks", Butterworth-Heinemann, 2003

[32] Kobielus, J. "New Federation Frontiers In IP Network Services", Publication: Business Communications Review. Date: Tuesday, August 1 2006

[33] C. Chapman, W. Emmerich, F. Galn, S. Clayman, A. Galis "Elastic Service Management in Computational Clouds", 12th IEEE/IFIP NOMS2010 / International Workshop on Cloud Management (CloudMan 2010) 19–23 April 2010, Osaka

[34] Strassner, J., Ó Foghlú, M., Donnelly, W. Agoulmine, N. "Beyond the Knowledge Plane: An Inference Plane to Support the Next Generation Internet", IEEE GIIS 2007, 2–6 July, 2007.

[35] Serrano, J.M. Strassner, J. and ÓFoghlú, M. "A Formal Approach for the Inference Plane Supporting Integrated Management Tasks in the Future Internet" 1st IFIP/IEEE ManFI International Workshop, In conjunction with 11th IFIP/IEEE IM2009, 1–5 June 2009, at Long Island, NY, USA

[36] Blumenthal, M., Clark, D. "Rethinking the design of the Internet: the end to end arguments vs. the brave new world", ACM Transactions on Internet Technology, Vol. 1, No. 1, Aug. 2001

[37] Bijan, P. et al. "Cautiously Approaching SWRL". 2006 http://www.mind swap.org/papers/CautiousSWRL.pdf.

[38] Mei, J., Boley, H. "Interpreting SWRL Rules in RDF Graphs". Electronic Notes in Theoretical Computer Science (Elsevier) (151): 53–69. 2006

[39] Neiger, G., Santoni, A., Leung, F., Rodgers D. and Uhlig, R.. "Intel Virtualization Technology: Software-only virtualization with the IA-32 and Itanium architectures", Intel Technology Journal, Volume 10 Issue 03, August 2006

[40] Cisco, VMWare. "DMZ Virtualization using VMware vSphere 4 and the Cisco Nexus" 2009 www.vmware.com/files/pdf/dmz-vsphere-nexus-wp.pdf

[41] Host your web site in the cloud, Jeff Barr, Sitepoint, 2010, ISBN 978–0-9805768-3-2

[42] V. Holub, T. Parsons, P. O'Sullivan, and J. Murphy. Run-time correlation engine for system monitoring and testing. In ICAC-INDST '09: Proceedings of the 6th international conference industry session on Autonomic computing and communications industry session, pages 9–18, New York, NY, USA, 2009. ACM.

[43] J. Shao, H. Wei, Q. Wang, and H. Mei. A runtime model based monitoring approach for cloud. In Cloud Computing (CLOUD), 2010 IEEE 3rd International

[44] The Real Meaning of Cloud Security Revealed, Online access Monday, May 04, 2009 http://devcentral.f5.com/weblogs/macvittie/archive/2009/05/04/the-real-meaning-of-cloud-security-revealed.aspx

[45] Serrano, J. Martin., Serrat, Joan., Strassner, John., Ó FoghlÚ, Mícheál. "Facilitating Autonomic Management for Service Provisioning using Ontology-Based Functions & Semantic Control" 3rd IEEE International Workshop on Broadband Convergence Networks (BcN 2008) in Conjuction with IEEE/IFIP NOMS 2008: Pervasive Management for Ubiquitous Network and Services. 07-11 April 2008, Salvador de Bahia, Brazil.

[46] Danh Le Phuoc, Martin Serrano, Hoan Nguyen Mau Quoc and Manfred Hauswirth, "A complete stack for building Web-of-Things Applications", submitted to the IEEE Computer Systems Journal, Special issue on the Web-of-Things 2013.

[47] John Soldatos, Martin Serrano, Manfred Hauswirth "Convergence of Utility Computing with the Internet-of-Things" IEEE 2012 Sixth International Conference on Innovative Mobile and Internet Services in Ubiquitous Computing (IMIS 2012), (ISBN: 978-1-4673-1328-5. DOI: 10.1109/IMIS.2012.135),

[48] M. Compton, P. Barnaghi, L. Bermudez, R. Garca Castro, O. Corcho, S. Cox, J. Graybeal, M. Hauswirth, C. Henson, A. Herzog, V. Huang, K. Janowicz, W. D. Kelsey, D. Le Phuoc, L. Lefort, M. Leggieri, H. Neuhaus, A. Nikolov, K. Page, A. Passant, A. Sheth, and K. Taylor. "The SSN Ontology of the Semantic Sensor Networks" Incubator Group Technical Report. JWS, 2012.

Index